The 2nd International Conference on Advances in Mechanical Engineering

The 2nd International Conference on Advances in Mechanical Engineering

Editors

Muhammad Mahabat Khan
Mohammad Javed Hyder
Muhammad Irfan
Manzar Masud

MDPI • Basel • Beijing • Wuhan • Barcelona • Belgrade

Editors
Muhammad Mahabat Khan
Capital University of Science and Technology
Pakistan

Mohammad Javed Hyder
Capital University of Science and Technology
Pakistan

Muhammad Irfan
Capital University of Science and Technology
Pakistan

Manzar Masud
Capital University of Science and Technology
Pakistan

Editorial Office
MDPI
St. Alban-Anlage 66
4052 Basel, Switzerland

This is a reprint of Proceedings published online in the open access journal *Engineering Proceedings* (ISSN 2673-4591) in 2022 (available at: https://www.mdpi.com/2673-4591/23/1).

For citation purposes, cite each article independently as indicated on the article page online and as indicated below:

LastName, A.A.; LastName, B.B.; LastName, C.C. Article Title. *Journal Name* **Year**, *Article Number*, Page Range.

ISBN 978-3-0365-5737-3 (Pbk)
ISBN 978-3-0365-5738-0 (PDF)

Contents

About the Editors

Muhammad Mahabat Khan

Muhammad Mahabat Khan (Dr.) is an Associate Professor and Head of the Mechanical Engineering Department at Capital University of Science and Technology (CUST), Islamabad, Pakistan. He obtained a Ph.D. degree in mechanical engineering with a specialization in Computation Fluid Dynamics from Ecole Centrale de Lyon, France. He served as an Advanced Development Engineer at the Systems Engineering division in Continental Automotive, France. He also worked as a Research Scholar in the Department of Mechanical Engineering, University of Leeds, UK. He has authored more than 30 WOS-indexed journal articles and serves as a reviewer for several international journals. His research interests include: CFD, engine sprays, heat-transfer augmentation, latent thermal energy storage systems, phase-change materials, fluidic oscillators, renewable energy, multi-phase flows, and turbulent flows.

Mohammad Javed Hyder

Mohammad Javed Hyder (Dr.) is a full professor at the Capital University of Science and Technology (CUST), Islamabad, Pakistan. His field of research focuses on renewable energy, Stirling engine design and development, mechanical system design, turbomachinery, nechanical behaviour of materials, and computational engineering. He has published around 30 articles in the journals and presented more than 35 research papers in the conferences. He has supervised 5 Doctoral theses and 175 Master's theses. He has organized various international conferences. He has teaching experience of about 40 years. He has taught more than 50 subjects in the field of mechanical engineering, material behaviour, computer science and engineering management to graduate and undergraduate students. He has vast experience in management as Dean of Faculty of Engineering, Head of Department of Mechanical Engineering, Head Department of Information and Public Relation and Director of Offices of Research, Innovation and Commercialization (ORICs).

Muhammad Irfan

Muhammad Irfan (Dr.) is an Assistant Professor of Mechanical Engineering at Capital University of Science and Technology (CUST), Islamabad. He obtained his PhD from Koc University Turkey, during which he worked on the development of a front-tracking multiphase phase change solver for the evaporation and combustion of fuel droplets. Chemkin package was integrated with the phase change solver to handle the chemical kinetics part. His current research interests include phase-change material-based energy-storage devices, performance enhancements of the photovoltaic panels, multiphase flows and ejector refrigeration system. He has published more than 10 articles in reputable international journals.

Manzar Masud

Manzar Masud is a lecturer at the Department of Mechanical Engineering, Capital University of Science and Technology (CUST), Islamabad, Pakistan. He received his BS degree in aerospace engineering from Institute of Space Technology, Pakistan and an MS degree in Mechanical Engineering from HITEC University, Pakistan. At present, he is a PhD candidate in the Department of Mechanical Engineering at the School of Mechanical and Manufacturing Engineering, National University of Science and Technology, Pakistan. His research interests include computational mechanics, the material characterization of composite materials, mechanical behavior of bio-hybrid composite materials and solid mechanics.

Preface to "The 2nd International Conference on Advances in Mechanical Engineering"

International Conference on Advances in Mechanical Engineering (ICAME) is organized by the Mechanical Engineering Department of Capital University of Science & Technology (CUST), Islamabad, Pakistan. ICAME is an annual event and usually takes place in the last week of August. ICAME aims to attract engineers, scientists and research scholars from leading academic institutes and industry around the world to disseminate the latest research and innovations. ICAME also encourages participants to present novel ideas and sustainable solutions to tackle local and global engineering challenges. ICAME covers all the major areas of the Mechanical Engineering and Engineering Management.

The second International Conference on Advances in Mechanical Engineering (ICAME-22) was held on August 25, 2022. ICAME-22 published 36 articles after a strict peer review process. More than 100 registered participants attended the conference from industry and academia. Five International keynote speakers joined the conference virtually, namely Prof. Alexey Burluka, Northumbria University; Prof. Amir Farooq, King Abdullah University of Science and Technology; Prof. Neyara Radwan, Suez Canal University; Prof. Lei Yue, Guangzhou University; Prof. Sam Yu, Vice President Tinbowah Investment China. The Conference proceedings were published as a Special Edition of Engineering Proceeding, Volume 23, 2022.

Muhammad Mahabat Khan, Mohammad Javed Hyder, Muhammad Irfan, and Manzar Masud
Editors

engineering proceedings

MDPI

Editorial

Statement of Peer Review †

Muhammad Mahabat Khan , Mohammad Javed Hyder *, Muhammad Irfan and Manzar Masud

Department of Mechanical Engineering, Capital University of Science and Technology (CUST), Islamabad 44000, Pakistan

* Correspondence: javed.hyder@cust.edu.pk; Tel.: +92-51-111555666 (ext. 282)

† Presented at the 2nd International Conference on Advances in Mechanical Engineering (ICAME-22), Islamabad, Pakistan, 25 August 2022.

In submitting conference proceedings to *Engineering Proceedings*, the volume editors of the proceedings certify to the publisher that all papers published in this volume have been subjected to peer review administered by the volume editors. Reviews were conducted by expert referees to the professional and scientific standards expected of a proceedings journal.

- Type of peer review: single-blind; double-blind; triple-blind; open; other (please describe): single-blind
- Conference submission management system: Online Portal
- Number of submissions sent for review: 45
- Number of submissions accepted: 36
- Acceptance rate (number of submissions accepted/number of submissions received): 80%
- Average number of reviews per paper: 2
- Total number of reviewers involved: 21.

check for updates

Citation: Khan, M.M.; Hyder, M.J.; Irfan, M.; Masud, M. Statement of Peer Review. *Eng. Proc.* **2022**, *23*, 37. https://doi.org/10.3390/engproc2022023037

Published: 11 October 2022

engineering proceedings

MDPI

Proceeding Paper

Study of Steam Jet Condensation for CD Spray Nozzle Exhausting into Quiescent Water †

Haseeb Afzal [1,*], Ajmal Shah [2], Abdul Quddus [1], Noman Arif Khan [1], Shumail Hassan [1], Muhammad Khawar Ayub [3] and Mazhar Iqbal [1]

[1] Department of Mechanical Engineering, Pakistan Institute of Engineering and Applied Sciences, PIEAS, Islamabad 45650, Pakistan
[2] Center for Mathematical Sciences, Pakistan Institute of Engineering and Applied Sciences, PIEAS, Islamabad 45650, Pakistan
[3] Theoretical Physics Division, Pakistan Institute of Nuclear Science & Technology (PINSTECH), Islamabad 45650, Pakistan
* Correspondence: haseebafzal36@gmail.com
† Presented at the 2nd International Conference on Advances in Mechanical Engineering (ICAME-22), Islamabad, Pakistan, 25 August 2022.

Abstract: Direct contact condensation (DCC) has achieved a well-known significance because of exceptional reasons such as efficient heat and mass transfer characteristics. The current experimental investigation involves considering the steam cavity shape characteristics with varying steam pressure, when the saturated steam is condensed into the one-phase water atmosphere using a converging-diverging (CD) nozzle. The results indicate the four different shapes of steam jet (oscillatory, conical, ellipsoidal and double expansion–contraction). It is observed that the penetration length and the maximum expansion ratio increase with the increase in steam saturated pressure and are found in the range of 1.8–2.8 and 1–1.13, respectively. Furthermore, the current results for jet length are compared with previously developed jet length predicting models which are found to be in good agreement.

Keywords: converging diverging nozzle; direct contact condensation; multiphase flows; steam cavity length; steam cavity shape

Citation: Afzal, H.; Shah, A.; Quddus, A.; Khan, N.A.; Hassan, S.; Ayub, M.K.; Iqbal, M. Study of Steam Jet Condensation for CD Spray Nozzle Exhausting into Quiescent Water. *Eng. Proc.* **2022**, *23*, 6. https://doi.org/10.3390/engproc2022023006

Academic Editors: Mahabat Khan, M. Javed Hyder, Muhammad Irfan and Manzar Masud

Published: 20 September 2022

1. Introduction

DCC occurs when the vapors are brought directly in contact with the cold liquid [1]. Kerney et al. [2] studied steam plume length whose correlation was developed as a function of subcooled water temperature, steam mass velocity and nozzle cavity diameter. Kerney's model was later modified by Weimer et al. [3] using turbulent entrainment and variable density theories. The calculations for chugging, oscillatory and stratified flows were executed by Aya et al. [4].

Later on, Chun et al. [5] presented the qualitative regime map of the saturated steam using both vertical and horizontal nozzles with the help of a total of 346 experiments. Kim et al. [6] attempted to measure the dimensionless length and expansion ratios of steam cavity for various steam conditions. Due to the unavailability of DCC involving supersonic steam jet, Wu et al. [7] presented the jet shape of the submerged supersonic nozzle. Shah et al. [8] examined the DCC of steam discharging into the liquid water atmosphere using the lab scale model of steam jet pump.

The current research aims to study steam plume parameters such as plume shape, dimensionless penetration length and maximum expansion ratio when the saturated steam is exhausted into the quiescent water tank via DCC. The outcomes would be significant in safe and economic design of DCC based equipment as well as in the validation of CFD results.

2. Materials and Methods

The experimental setup utilized in the current research is shown in Figure 1. This setup consists of electric boiler, surge tank, water tank, CD nozzle, instrumentation, valves, mobile measuring probe and high-speed camera. The steam generator, which is most crucial component of the setup, is used to supply saturated steam with the quality of ~99% having a peak value of operating pressure of 8 bar. The steam is injected horizontally via the CD nozzle. A rectangular water tank with dimensions of 1.2 m × 1 m × 1 m is used for the water storage. The captured images are processed using MATLAB to calculate the jet parameters. The operating and geometric conditions of the present experimental investigation are given in Table 1.

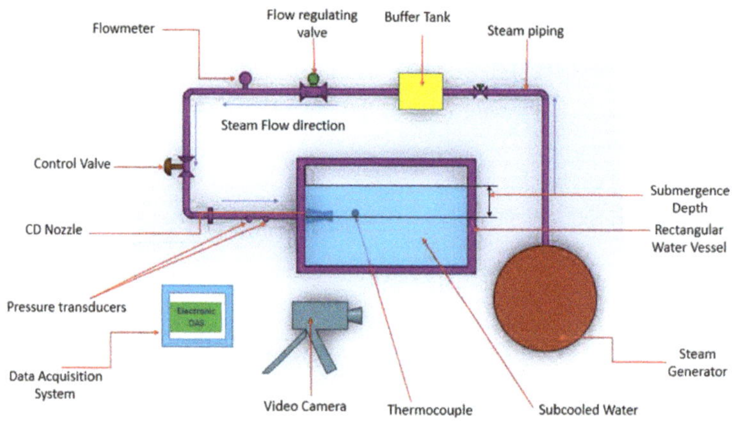

Figure 1. DCC experimental facility.

Table 1. Operating conditions.

Parameters	Value/Range
Steam pressure (Absolute)	1.5–4.5 bar
Water temperature, Tw	35 °C
Nozzle inlet diameter	10 mm
Throat diameter of nozzle	5 mm
Nozzle submergence depth	200 mm

3. Results and Discussion

3.1. Influence of Steam Saturated Pressure on the Jet Shapes

The alteration of steam cavity shapes with the steam saturated pressure at the constant temperature of 35 °C is shown in Figure 2. At the low pressures of 1.5 bar and 2 bar, (a) and (b) show the unstable steam jet shapes and oscillations take place at the steam water interface. This phenomenon is called condensation oscillations. At the pressures of 2.5 bar and 3 bar, the shape of steam plume is found to be conical as shown in (c) and (d). This type of shape is formed if the pressure at the exit of the nozzle is kept below the atmospheric pressure, hence resulting in the compression of plume.

However, as the pressure is further increased, an ellipsoidal shape was observed as shown in (e) and (f). The formation of this type of shape is associated with the theory of expansion and contraction. If the pressure is increased further to 4.5 bar, the plume manifests a double expansion–contraction shape which is shown in (g). The formation of this shape is associated with the fact that as the pressure elevates further at the constant temperature, momentum is imparted along with the transfer of thermal energy to the water.

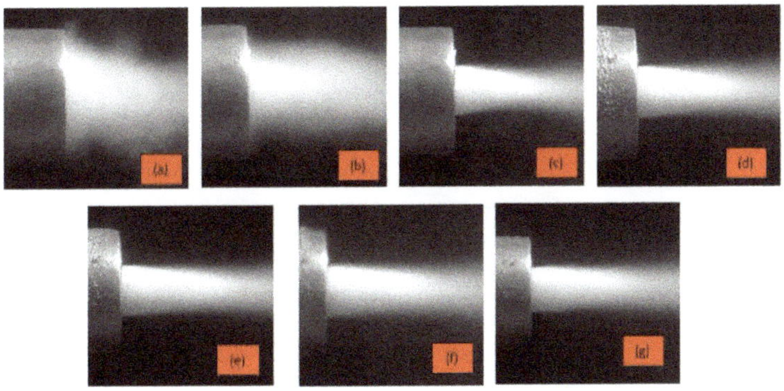

Figure 2. Steam pressure effect on jet shapes at 35 °C: (**a**) 1.5, (**b**) 2, (**c**) 2.5, (**d**) 3, (**e**) 3.5, (**f**) 4, (**g**) 4.5.

3.2. Influence of Steam Saturated Pressure on the Dimensionless Penetration Length

The variation of penetration length with the steam pressure at a temperature of 35 °C is shown as the black line in Figure 3. The parameter is taken in the dimensionless form by dividing the penetration length over the nozzle exit diameter. It is clear from the trend that as the steam pressure is increased, the dimensionless penetration length goes on increasing. This variation is obvious because at higher steam pressure, higher momentum is imparted, and the input content of heat energy also rises. The values of dimensionless penetration were found in the range of 1.8–2.8 which is in well accordance with the range of previous researchers at the low temperatures.

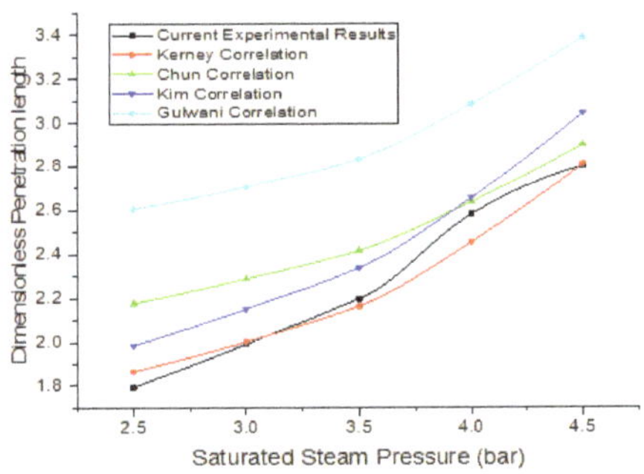

Figure 3. Comparison of Current Experimental Results with the Predicted Data.

The dimensionless penetration length is also predicted using the various correlations given in the literature for the comparison purposes. Overall, all the correlations determined the much reasonable agreement lying within the −30% and +5% as a whole with the absolute average deviation of 10.2%. The inflection point is observed in the current data because of the variation of shape from ellipsoidal to double-expansion contraction.

3.3. Influence of Steam Saturated Pressure on the Expansion Ratio

It has been found that as the saturated steam pressure increases, the maximum expansion ratio also increases. The reason behind the increasing trend of maximum expansion ratio with the steam pressure can be explained on the basis of higher momentum transfer which expands the steam in the radial direction. At the low pressures, the maximum expansion ratio was simply 1 because of the convergent shape; however, it increases as the plume becomes ellipsoidal at higher pressures. It was found to be maximum at the maximum pressure of 4.5 bar where the double expansion–contraction shape was found. The maximum expansion ratio was obtained in the range of 1–1.133 which is in accordance with the previous studies at lower temperatures.

4. Conclusions

The following are some of the major conclusions obtained after the current investigation:

1. A total of four different steam plume shapes were observed including oscillatory, conical, ellipsoidal and double contraction–expansion shape at the relatively lower temperature of 35 °C within a pressure range of 1.5–4.5 bar.
2. The dimensionless penetration length increases with the increase in steam pressure and was found in the range of 1.8–2.8.
3. Among the four correlations employed for the prediction of penetration length, Kerney's correlation agreed the most with the experimental data while the maximum discrepancy found throughout was between the data and Gulwani's correlation.
4. The maximum expansion ratio also increases with the increase in steam pressure and was found in the range of 1–1.13.

Author Contributions: Conceptualization, H.A., A.Q. and A.S.; methodology, H.A. and M.I.; software, H.A., A.Q. and N.A.K.; validation, H.A. and N.A.K.; formal analysis, H.A. and S.H.; investigation, H.A. and M.I.; data curation, M.K.A. and S.H.; writing—original draft preparation, H.A.; writing—review and editing, H.A. and M.I.; supervision, M.I.; project administration, H.A. All authors have read and agreed to the published version of the manuscript.

Funding: This research received no external funding.

Acknowledgments: The authors acknowledge financial, technical, and administrative support provided by Pakistan Institute of Engineering and Applied Sciences for the present study.

Conflicts of Interest: The authors declare no conflict of interest.

References

1. Wu, X.-Z.; Yan, J.-J.; Pan, D.-D.; Liu, G.-Y.; Li, W.-J. Condensation Regime Diagram for Supersonic/Sonic Steam Jet in Subcooled Water. *Nucl. Eng. Des.* **2009**, *239*, 3142–3150. [CrossRef]
2. Kerney, P.J.; Faeth, G.M.; Olson, D.R. Penetration Characteristics of a Submerged Steam Jet. *AIChE J.* **1972**, *18*, 548–553. [CrossRef]
3. Weimer, J.C.; Faeth, G.M.; Olson, D.R. Penetration of Vapor Jets Submerged in Subcooled Liquids. *AIChE J.* **1973**, *19*, 552–558. [CrossRef]
4. Aya, I.; Nariai, H. Evaluation of Heat-Transfer Coefficient at Direct-Contact Condensation of Cold Water and Steam. *Nucl. Eng. Des.* **1991**, *131*, 17–24. [CrossRef]
5. Chun, M.-H.; Kim, Y.-S.; Park, J.-W. An Investigation of Direct Condensation of Steam Jet in Subcooled Water. *Int. Commun. Heat Mass Transf.* **1996**, *23*, 947–958. [CrossRef]
6. Kim, H.Y.; Bae, Y.Y.; Song, C.H.; Park, J.K.; Choi, S.M. Experimental Study on Stable Steam Condensation in a Quenching Tank. *Int. J. Energy Res.* **2001**, *25*, 239–252. [CrossRef]
7. Wu, X.-Z.; Yan, J.-J.; Shao, S.-F.; Cao, Y.; Liu, J.-P. Experimental Study on the Condensation of Supersonic Steam Jet Submerged in Quiescent Subcooled Water: Steam Plume Shape and Heat Transfer. *Int. J. Multiph. Flow* **2007**, *33*, 1296–1307. [CrossRef]
8. Shah, A.; Chughtai, I.R.; Inayat, M.H. Experimental and Numerical Analysis of Steam Jet Pump. *Int. J. Multiph. Flow* **2011**, *37*, 1305–1314. [CrossRef]

engineering proceedings

Proceeding Paper

Experimental Study on Steam Cavity Characteristics for Swirled Flow Nozzle Exhausting into Quiescent Water †

Abdul Quddus [1,*], Ajmal Shah [2], Kamran Rasheed Qureshi [1], Muhammad Khawar Ayub [3], Mazhar Iqbal [1], Noman Arif Khan [1], Haseeb Afzal [1] and Shumail Hassan [1]

1 Department of Mechanical Engineering, Pakistan Institute of Engineering and Applied Sciences, PIEAS, Islamabad 45650, Pakistan
2 Center for Mathematical Sciences, Pakistan Institute of Engineering and Applied Sciences, PIEAS, Islamabad 45650, Pakistan
3 Theoretical Physics Division, PINSTECH, Islamabad 44000, Pakistan
* Correspondence: engquddus613@yahoo.com
† Presented at the 2nd International Conference on Advances in Mechanical Engineering (ICAME-22), Islamabad, Pakistan, 25 August 2022.

Abstract: The steam–water direct contact condensation (DCC) process is commonly observed in various industries due to its fast heat and mass exchange characteristics. This study investigates steam plume characteristics by experimentally condensing the steam jet issuing from a swirled flow spray nozzle into stagnant subcooled water. On the basis of high-speed imaging, the effects of subcooled water temperatures on the cavity shape, its length, and maximum expansion ratio were explored. The existence of three distinct cavity shapes (ellipsoidal, double expansion–contraction and divergent) were identified. The dimensionless steam cavity penetration length and maximum expansion ratio were found to be in the range of 6.28–11.5 and 1.71–3.06, respectively. The results indicate that with the rise in water temperature, plume length and maximum expansion ratio increase.

Keywords: multiphase flow; plume cavity; swirl flow; direct contact condensation; fluid dynamics

Citation: Quddus, A.; Shah, A.; Qureshi, K.R.; Ayub, M.K.; Iqbal, M.; Khan, N.A.; Afzal, H.; Hassan, S. Experimental Study on Steam Cavity Characteristics for Swirled Flow Nozzle Exhausting into Quiescent Water. *Eng. Proc.* **2022**, *23*, 28. https://doi.org/10.3390/engproc2022023028

Academic Editors: Mahabat Khan, Muhammad Javed Hyder, Muhammad Irfan and Manzar Masud

Published: 22 September 2022

1. Introduction

The phenomenon of steam-water DCC of steam has become of paramount interest in the industrial sector due to the high exchange rate of heat and mass at the interface [1]. Its typical applications in industry includes direct mixing heaters, thermal degasification systems, steam injectors, and nuclear reactor safety systems [2–4].

In the published literature, scholars mainly give stress to the study of steam jet characteristics including steam cavity shapes, cavity length, temperature profiles, heat exchange rate, and condensation oscillations [1,5,6]. Recently, researchers have considered the steam cavity shape as a key parameter during submerged jet condensing with water. It is also strongly linked with the above said research areas [4–8]. Additionally, it provides guidelines in the design and functionality of the DCC based systems [4,7]. Hence, the topic of the steam jet characteristics have become attractive in the field of DCC.

Previously published literature indicates that steam cavity characteristics depend upon the operating and geometric conditions [6–8]. Kerney's study [8] was the first a predicting model for steam cavity axial size was formulated. Weimer et al. [9] modified Karney's model by including phasic densities effect.

Shah et al. [7] pointed out that the steam cavity shape is dependent on nozzle geometric parameters. Zhang and his group [10] studied the steam jet features for double hole nozzle under stable flow conditions. Quddus et al. [11,12] studied the steam jet features from the beveled steam spray nozzles. They also developed a predicting model for steam jet length by including the effects of nozzle exit bevel angle. In their study, they explored the physics of the jet shapes for various beveled nozzles.

The purpose of the present work is to investigate the characteristics of steam cavity shapes for swirled flow spray nozzles discharging into stagnant water. A high speed imaging technique was used for capturing the steam cavity shapes. The present findings are believed to be helpful for gaining more insight into steam-water DCC systems and understanding the associated problems.

2. Experimental Setup and Methodology

The details of the experimental facility and methodology are the same as presented in our previous work [11,12]. The only difference is the type of steam nozzle used. Here, as mentioned earlier, a steam swirled flow spray nozzle has been used which is shown in Figure 1.

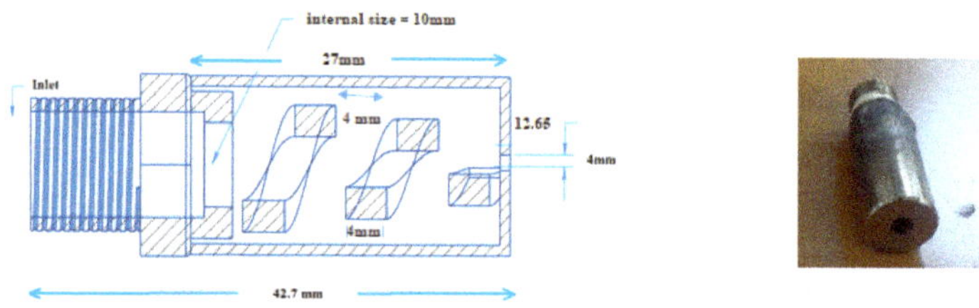

(a) (b)

Figure 1. (**a**) Section view of steam swirled flow spray nozzle. (**b**) Fabricated nozzle.

3. Results and Discussion

3.1. *Effect on Steam Cavity Shape*

This section presents the variation in cavity shapes under the action of the changing tank water temperature for steam issuing from a swirled flow spray nozzle exhausting into stagnant water at constant steam pressure, P_s = 500 KPa.

Figure 2a–e illustrates the steam cavity shape observed at various water temperatures i.e., T_W = 30, 40, 50, 60, and 70 °C, respectively. From this figure, it can be seen that at T_W = 30 °C, the ellipsoidal cavity shape exists as shown in Figure 2a. The formation of such a cavity shape can be explained by the steam jet becoming under expanded at elevated steam pressures. From expansion-contraction theory, such cavities are expanded as the steam is ejected out from the nozzle at the exit location. When T_W = 50 and 60 °C, condensation capabilities of cooling medium decrease. To dissipate the extra input of steam thermal energy, additional expansion of the plume occurs by increasing the surface area of the cavity. The cavity is again compressed by nearby water which is at relatively high pressure. The plume thus becomes inflexed which finally condenses to ambient water conditions. Consequently, the double expansion–contraction shaped cavity is formulated, as presented in Figure 2b,c. When T_W = 60 and 70 °C, there is continuous expansion in the cavity which transforms the shape of the plume into a divergent jet as illustrated in Figure 2d,e. For such shapes, the losses for heat exchange capabilities become large due to escaping of steam bubbles which carry the heat with them.

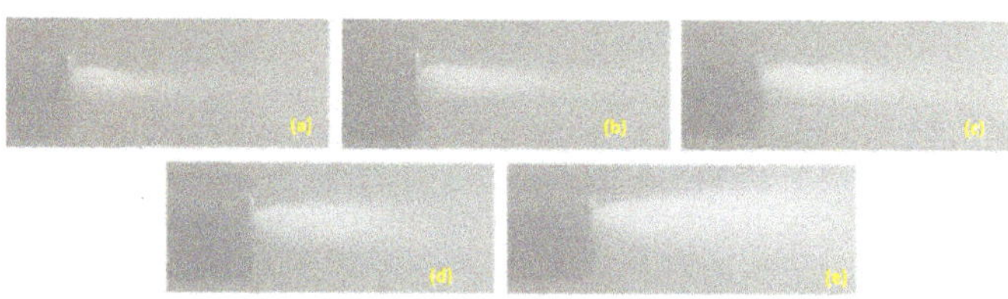

Figure 2. Steam cavity shape observed at P_s= 500 kPa and at T_w = (**a**) 30 °C, (**b**) 40 °C, (**c**) 50 °C, (**d**) 60 °C, and (**e**) 70 °C.

3.2. *Effect on Steam Cavity Length*

Figure 3a describes the effects of liquid phase temperature on steam cavity length at constant steam pressure, i.e., 500 kPa. The plume length was measured at five water temperatures, i.e., T_W = 30, 40, 50, 60 and 70 °C. From the findings, it can be concluded that with the decrease in degree of subcooling of water, the steam cavity length prolongs in axial dimensions. This fact can be explained as follows: as the degree of subcooling of water decreases, its temperature rises, and its corresponding condensing capability is reduced. The heat transfer area of the cavity is increased to thermally balance the input heat. Consequently, the penetration length of the steam jet increases.

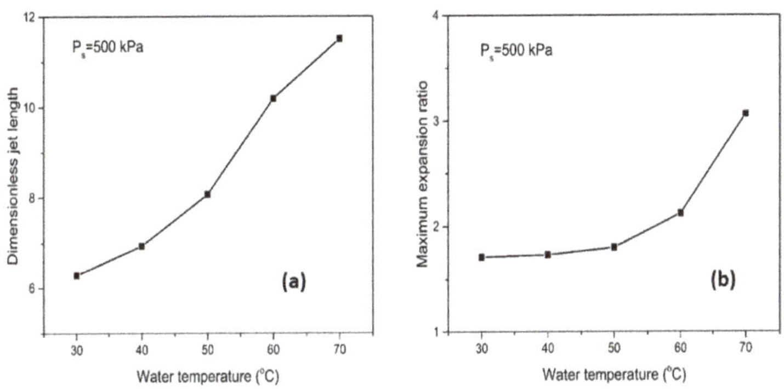

Figure 3. Effect of tank water temperature on (**a**) cavity dimensionless length, and (**b**) maximum expan-sion ratio.

3.3. *Effect on Steam Cavity Maximum Expansion Ratio*

Figure 3b shows the variation of the steam jet maximum expansion ratio with water temperature at constant steam pressure P_S = 500 kPa. The maximum expansion ratio has been calculated at various degrees of subcooling of water by changing water temperature (i.e., T_W = 30, 40, 50, 60 and 70 °C). It was observed that with the decrease in degree of subcooling, the steam jet maximum expansion ratio increases. This can be explained by the condensation capacity of the water reducing with increase in water temperature. Moreover, the area across which the jet exchanges its heat with the liquid phase is increased to thermally balance the heat. The increase in area is both from the axial and radial growth of the jet. Consequently, the radial growth of the cavity diameter increases the maximum expansion ratio.

4. Conclusions

In the current study, a steam jet from a swirled flow spray nozzle was condensed experimentally into stagnant subcooled water. A high speed imaging technique was utilized to explore the steam cavity characteristics. The effect of water temperature (range, 30–70 °C) on cavity parameters at steam inlet pressure 500 kPa was studied. The following main conclusions were obtained.

1. A total of three different plume shapes exist for the above-described flow conditions: ellipsoidal, double expansion–contraction, and divergent.
2. At lower water temperature, an ellipsoidal plume shape was observed. At intermediate water temperature, the double expansion–contraction shape was recorded, while the divergent cavity presented at high water temperature.
3. The steam cavity length and maximum expansion ratio were found to be in the range of 6.28–11.5 and 1.71–3.06, respectively.
4. With the increase in the subcooled water temperature, both the steam cavity length and maximum expansion ratio increased.

Author Contributions: A.Q.: Conceptualization, methodology, validation, investigation, data curation, original draft preparation; A.S.: Conceptualization, writing—review and editing, supervision, project administration; K.R.Q.: investigation, supervision, project administration. M.K.A.: formal analysis, data curation, software; M.I.: investigation, project administration; N.A.K.: formal analysis, data curation; H.A.: formal analysis, data curation; S.H.: formal analysis, data curation. All authors have read and agreed to the published version of the manuscript.

Funding: The project was funded by Higher Education Commission of Pakistan under HEC indigenous PhD fellowship scheme (417-47877-2EG4-042).

Institutional Review Board Statement: Not applicable.

Informed Consent Statement: Not applicable.

Data Availability Statement: Data may be available on request.

Acknowledgments: The authors acknowledge the financial assistance from the Higher Education Commission, Pakistan for this work.

Conflicts of Interest: The authors declare no conflict of interest.

References

1. Qiu, B.; Yan, J.; Liu, J.; Chong, D.; Zhao, Q.; Wu, X. Experimental Investigation on the Second Dominant Frequency of Pressure Oscillation for Sonic Steam Jet in Subcooled Water. *Exp. Therm. Fluid Sci.* **2014**, *58*, 131–138. [CrossRef]
2. Shukla, S.K.; Samad, A.M.N.; Ghosh, S. CFD Simulation of Steam Condensation in a Subcooled Water Pool. *Therm. Sci. Eng. Prog.* **2017**, *2*, 80–86. [CrossRef]
3. Qiu, B.; Tang, S.; Yan, J.; Liu, J.; Chong, D.; Wu, X. Experimental Investigationon Pressure Oscillations Caused by Direct Contact Condensation of Sonic Steam Jet. *Exp. Therm. Fluid Sci.* **2014**, *52*, 270–277. [CrossRef]
4. Zhao, Q.; Hibiki, T. Review: Condensation Regime Maps of Steam Submerged Jet Condensation. *Prog. Nucl. Energy* **2018**, *107*, 31–47. [CrossRef]
5. Gulawani, S.S.; Joshi, J.B.; Shah, M.S.; Rama Prasad, C.S.; Shukla, D.S. CFD Analysis of Flow Pattern and Heat Transfer in Direct Contact Steam Condensation. *Chem. Eng. Sci.* **2006**, *61*, 5204–5220. [CrossRef]
6. Kim, H.J.; Bankoff, S.G.; Murphy, W.P. Local Heat Transfer Coefficients for Condensation in Stratified Counter currentSteam-Water Flows. *J. Heat Transf.* **1983**, *105*, 706–712. [CrossRef]
7. Shah, A.; Chughtai, I.R.; Inayat, M.H. Experimental Study of the Characteristics of Steam Jet Pumpand Effect of Mixing Section Lengthon Direct-Contact Condensation. *Int. J. Heat Mass Transf.* **2013**, *58*, 62–69. [CrossRef]
8. Kerney, P.J.; Faeth, G.M.; Olson, D.R. Penetration Characteristics of a Submerged Steam Jet. *AIChE J.* **1972**, *18*, 548–553. [CrossRef]
9. Weimer, J.C.; Faeth, G.M.; Olson, D.R. Penetration of Vapor Jets Submerged in Subcooled Liquids. *AIChE J.* **1973**, *19*, 552–558. [CrossRef]
10. Zhang, Y.; Feng, L.; Liu, L.; Fu, X.; Qiu, Z.; Yang, Y.; Wang, Z.; Yuan, Y.; Lu, D. Experimental Research on Double-Hole Steam Jets Dcc Heat Transfer Characteristics under Stable Condensation Pattern. *Int. J. Heat Mass Transf.* **2020**, *158*, 119881. [CrossRef]
11. Quddus, A.; Shah, A.; Qureshi, K.R.; Ayub, M.K.; Tahir, A.; Iqbal, M.; Waheed, K.; Kaleem, M.A. Study of Steam Jet Characteristics and Regime Maps for Bevelled Spray Nozzles Exhausting into Quiescent Water. *Int. J. Heat Mass Transf.* **2022**, *190*, 122780. [CrossRef]
12. Quddus, A.; Shah, A.; Rasheed Qureshi, K.; Khawar Ayub, M.; Tahir, A.; Iqbal, M.; Samee, A. Experimental Study on Steam Plume Shape Characteristics for Bevelled Spray Nozzles Exhausting into Quiescent Water. *Nucl. Eng. Des.* **2022**, *394*, 111824. [CrossRef]

engineering proceedings

MDPI

Proceeding Paper

Numerical Investigation of Different Configurations of Pin Fin Heat Sinks with and without PCM †

Hamza Fayyaz *, Abid Hussain and Talal Bin Irshad

Department of Mechanical Engineering, University of Engineering and Technology Taxila, Punjab 47050, Pakistan
* Correspondence: hamza.fayyaz@students.uettaxila.edu.pk
† Presented at the 2nd International Conference on Advances in Mechanical Engineering (ICAME-22), Islamabad, Pakistan, 25 August 2022.

Abstract: The miniaturization of electronic components leads to poor heat dissipation, performance, and reliability of the devices. To optimize heat transfer from electrical components, pin fins are the best choice due to their high thermal conductivity. Different configurations of square and triangular pin fins for heat transfer characteristics were numerically investigated in this study. Three-dimensional numerical simulations were carried out for two different configurations with heat fluxes of 0.82 kW/m^2 subjected to the base of the heat sink. The heat transfer coefficients among the numerical simulation study and experimentation results provided a good correlation. To evaluate the overall performance, it was indicated from both numerical and experimental results that the maximum square pin fin configuration reduced the base temperature by up to 17.7% and 19%, respectively. In comparison to other configurations, the square pin fin performed better because of its increased surface area.

Keywords: numerical simulation; electronic devices; pin fin heat sink; heat transfer coefficient

check for updates

Citation: Fayyaz, H.; Hussain, A.; Irshad, T.B. Numerical Investigation of Different Configurations of Pin Fin Heat Sinks with and without PCM. *Eng. Proc.* **2022**, *23*, 22. https://doi.org/10.3390/engproc2022023022

Academic Editors: Mahabat Khan, M. Javed Hyder, Muhammad Irfan and Manzar Masud

Published: 21 September 2022

1. Introduction

Recent advancements in electronics have led to an exponential rise in the temperature and heat transfer rates, which are affected due to their smaller sizes [1]. The main purpose of heat sinks is to cool electrical equipment. Pin- and plate-finned heat sinks are the two key forms that increase thermal performance. [2]. There are wide range of heat sink design geometries, although the most common pin fin designs are square, triangular, circular, and elliptic shapes. The performance of heat sink is significantly impacted by increasing the number and thickness of fins [3]. Phase change materials alter their phases by absorbing a huge amount of heat during melting, and releasing it during solidification [4]. Numerous studies have been already performed on PCM conductivity, and to overcome this problem, researchers tend to add fin geometries, composites metal, and ultrafine particles in addition to PCM [5].

This study focuses on the numerical model investigation of various arrangements of (square and triangular) pin fin heat sinks with and without PCM. Numerical simulations using COMSOL Multiphysics 5.5 with various parameters were studied on finned heat sinks. Variables such as heat transfer rate, base temperature reduction, and the maximum temperature of all configurations were analyzed.

2. Numerical Methodology and Materials

In this study, a fin model for thermal performance characteristics is presented. The model was generated in AutoCAD software and imported into COMSOL Multiphysics 5.5. Aluminum grade 2024 was selected for both configurations. Two arrangements of pin fin configurations (square and triangular) were studied in this research (Figure 1). The base plate of the pin fin configurations had a surface area of 114 × 114 mm^2, and the thickness of the base plate was 4 mm. There were 72 fins in total, which were extruded up to 25 mm above the base plate (Table 1).

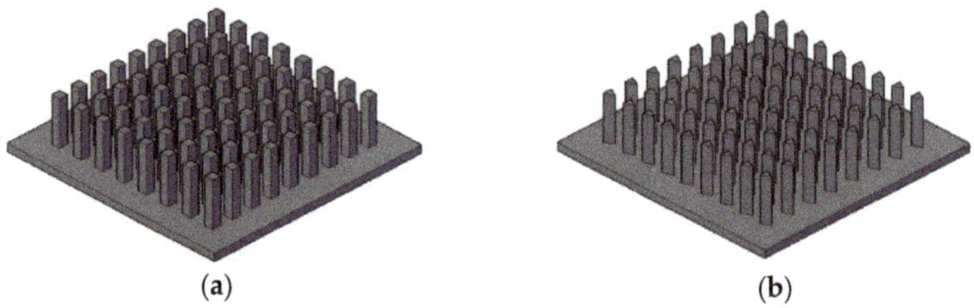

Figure 1. Pin fin heat sinks: (**a**) square; (**b**) triangular.

Table 1. Heat sink geometric properties.

Parameters	Dimensions (mm)
Fin Height	25
Fin Dimensions	4 × 4
Number of Fins	72
Heat Sink Length	114
Heat Sink Width	114

3. Experimental Setup

A comparison between heating and cooling of heat sink was conducted by employing different arrangements of pin fin heat sinks, i.e., square and triangular, with and without PCM. A power level of 10 W was used for this experiment and the heat sink discharged in the same way it charged, with the sidewalls and base insulated to enable heat flow in only one direction—from top to bottom—for thermal analysis. The experimental setup and exploded view of the heat sink assembly for this study are shown in Figures 2 and 3. Subsequently, the temperature contours of heat sink at different intervals of time are presented in Figure 4.

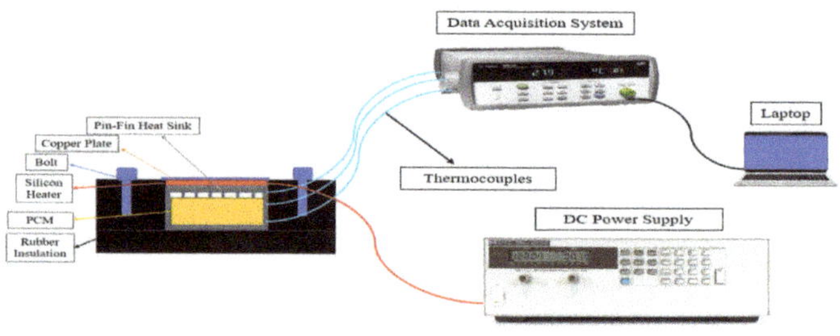

Figure 2. Experimental setup.

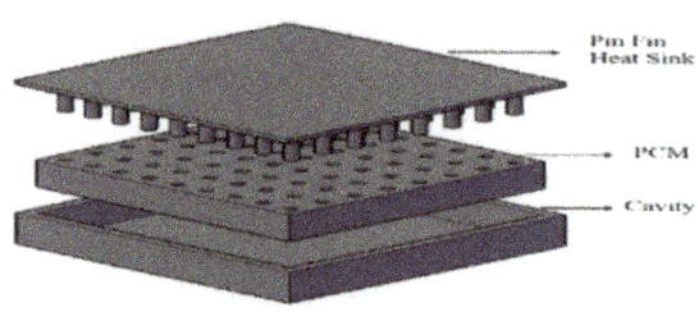

Figure 3. Exploded view of the assembly.

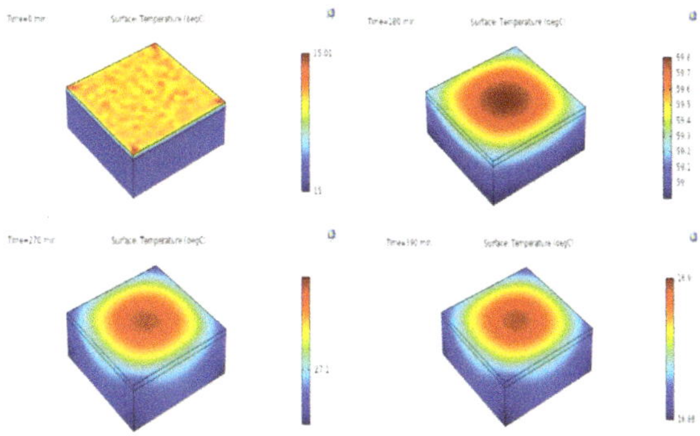

Figure 4. Temperature contours of a heat sink assembly at different intervals.

4. Results and Discussion

This study consisted of two phases: fins without PCM, and fins with PCM. The first set of simulations, performed for fins without PCM, and the second set of simulations, performed for fins with PCM, for both configurations, are shown in Figure 5.

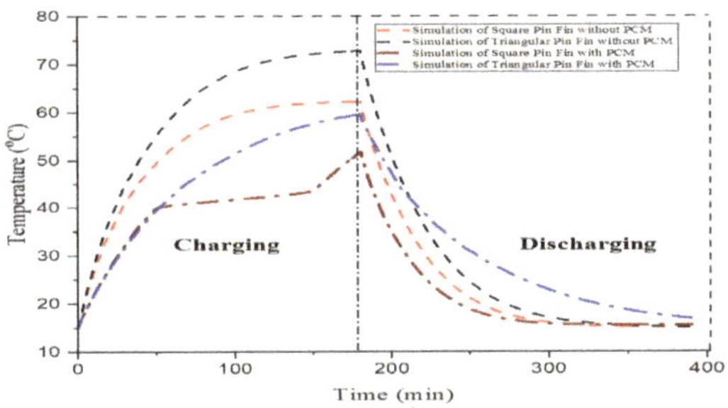

Figure 5. Simulation of square and triangular pin fins with and without PCM.

The reductions in base temperature are clearly seen by the in addition of PCM. Both configurations yielded the best results and maximum reduction in the base temperature of 17.7% with the square configuration, whereas the triangular configuration reduced the base temperature by up to 17.1%.

An experimental comparison of both configurations, with and without PCM, is shown in Figure 6. The maximum temperature reductions by square and triangular pin fins are 19% and 16%, respectively.

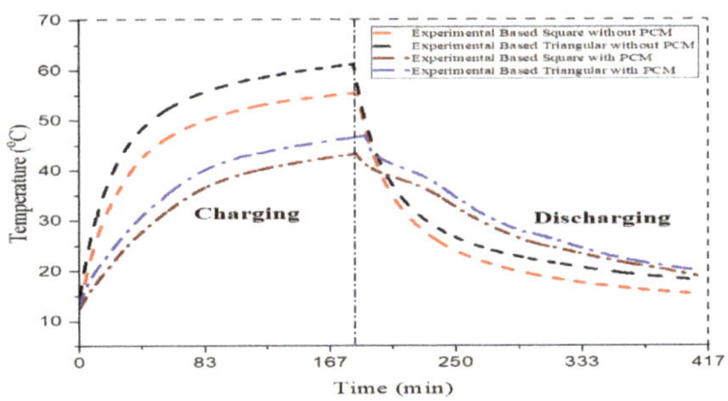

Figure 6. Experimental results of square and triangular pin fins.

Validation

Simulation results were compared with experimental data to validate the model shown in Figure 7. The observed error range was 2% for the pin fin with PCM and 3% for the pin fin without PCM, which was calculated using the formula: [Percentage error = Numerical Value—Experimental Value/Numerical Value], which was negligible. Overall, the numerical results showed the best results with the experimental data, indicating that the model was logical.

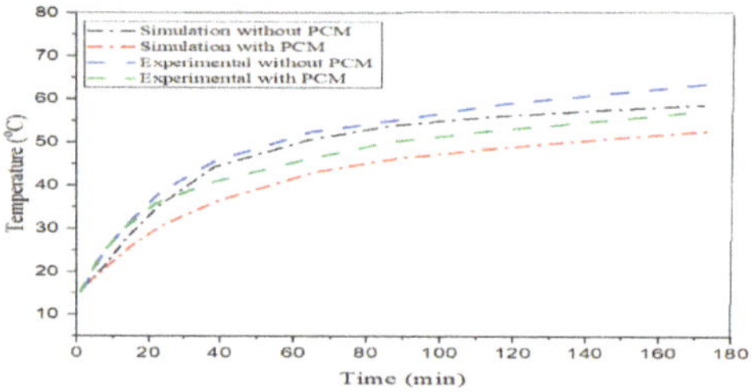

Figure 7. Validation of simulation results with experimental data.

5. Conclusions

This study focused on the reduction in base temperature in electronic applications. Numerical analyses of square and triangular pin fins provided the same results as the experimental data. The results concluded that adding PCM into the square configuration reduced the temperature by up to 17.7% as compared with a pin fin without PCM. The maximum peak temperature of the square pin fin without PCM was 62.1 °C, whereas the maximum peak temperature with PCM in the same configuration was 51.2 °C. The experimental results of the square configuration showed a maximum base temperature reduction of up to 19%. The present numerical simulation results of COMSOL Multiphysics were validated with an experimental model with the same pattern. Overall, the square configuration demonstrated better results for temperature reduction, due to the difference in surface area.

Author Contributions: H.F.: Conceptualization, methodology, validation, formal analysis, investigation, resources, data curation, writing—original draft preparation, T.B.I.: software, A.H.: writing—review and editing, supervision. All authors have read and agreed to the published version of the manuscript.

Funding: This research received no external funding.

Institutional Review Board Statement: Not applicable.

Informed Consent Statement: Not applicable.

Data Availability Statement: Not applicable.

Conflicts of Interest: The authors declare no conflict of interest.

References

1. Hithaish, D.; Saravanan, V.; Umesh, C.K.; Seetharamu, K.N. Thermal Management of Electronics: Numerical Investigation of Triangular Finned Heat Sink. *Therm. Sci. Eng. Prog.* **2022**, *30*, 101246. [CrossRef]
2. Jaffal, H.M.; Jebur, H.S.; Hussein, A.A. Numerical and Experimental Investigations on the Performance Characteristics for Different Shapes Pin Fin Heat Sink. *IJOCAAS* **2018**, *4*, 330–343.
3. Ali, S.A.; Al-Zuhairy, R.C.; Kubba, A.I. Three-Dimensional Numerical Simulation of Heat Transfer for Different Shapes of Heat Sinks. *J. Mech. Eng. Res. Dev.* **2021**, *44*, 345–362.

4. Reuben Raj, C.; Suresh, S.; Vasudevan, S.; Chandrasekar, M.; Kumar Singh, V.; Bhavsar, R.R. Thermal Performance of Nano-Enriched Form-Stable PCM Implanted in a Pin Finned Wall-Less Heat Sink for Thermal Management Application. *Energy Convers. Manag.* **2020**, *226*, 113466. [CrossRef]
5. Leong, K.Y.; Chew, S.P.; Gurunathan, B.A.; Ku Ahmad, K.Z.; Ong, H.C. An Experimental Approach to Investigate Thermal Performance of Paraffin Wax and 1-Hexadecanol Based Heat Sinks for Cooling of Electronic System. *Int. Commun. Heat Mass Transf.* **2019**, *109*, 104365. [CrossRef]

engineering proceedings

MDPI

Proceeding Paper

Acoustic Response of Fully Passive Airfoil under Gust †

Muhammad Arqam, Kashif Ayaz, Muhammad Ebrahem and Shehryar Manzoor *

Department of Mechanical Engineering, University of Engineering and Technology, Taxila 47070, Pakistan
* Correspondence: m.shehryar@uettaxila.edu.pk

† Presented at the 2nd International Conference on Advances in Mechanical Engineering (ICAME-22), Islamabad, Pakistan, 25 August 2022.

Abstract: Acoustic response from a freely responding symmetric airfoil subjected to gust is investigated in a two-dimensional numerical environment. Gust model is superimposed on the inlet velocity up till the critical flutter velocity. Second order transient formulation, $k-\omega$ turbulence model and dynamic meshing technique were adopted. By employing the Ffowcs Williams and Hawkings (FW-H) acoustic methodology, the acoustic signature generated by the airfoil for the range of velocities ($0.85 \leq U/U_c \leq 1$ near the critical flutter velocity is quantified over a range of acoustic receivers in the surrounding of the airfoil. Sound pressure levels (SPLs) are determined, and directionalities have been studied. It is revealed that the distribution of sound pressure level at the exciting frequency is affected by the gust profile. Scales of these sound pressure levels, however, relied on the Reynolds number and the dynamics of the system.

Keywords: gust response; flow noise; aeroacoustics; fully passive airfoil

Citation: Arqam, M.; Ayaz, K.; Ebrahem, M.; Manzoor, S. Acoustic Response of Fully Passive Airfoil under Gust. *Eng. Proc.* **2022**, *23*, 36. https://doi.org/10.3390/engproc2022023036

Academic Editors: Mahabat Khan, M. Javed Hyder, Muhammad Irfan and Manzar Masud

Published: 27 September 2022

1. Introduction

The interior noise and airframe noise have been explored extensively in the recent past. There is an ever-increasing urgency to mitigate the influence of aerospace productions on the environment with the European commission calling for noise reduction [1], which has resulted in the development of systems capable of meeting noise certification needs.

The flow of fluid over stiffened structures is known to yield sound, as well as disturbs and sustains vibrations in the structure. When a body moves in a nonuniform fluid flow, the contact between the body and the unsteady fluid harvests pressure fluctuations on the body surface resulting in the noise propagated to the far-field. This investigation has numerous applications in the design and mitigation of microaerial vehicles (MAVs) as well as aircraft structural noise.

Métivier et al. [2] performed a series of investigations experimentally on an unrestricted pitching wing (chord length = 0.156 m) at Reynolds numbers of 5.10^4–1.10^5 and revealed that the airfoil became unstable, exhibiting limit-cycle oscillations. Wind tunnel studies using a pitching and plunging airfoil in the previous Reynolds number range have also been performed [3]. The consequence of laminar separation on the flapping airfoil has been considered experimentally and numerically by [2,4]. These research studies confirm that the separation of the boundary layer in the laminar region at the trailing edge is accountable for pitching oscillations.

2. Governing Equations

The current effort utilised the two-dimensional unsteady Reynolds averaged Navier–Stokes (URANS) methodology improved with a transitional turbulent solver for the flow calculations ($k-\omega$ Shear Stress Transport). Moreover, the Ffowcs Williams and Hawkings's (FW-H) method was employed for the aeroacoustics calculations. The model was selected as it was not computationally expensive and was apt for predicting the tonal noise from the flow and the interface of the flow with nonpermeable contours.

2.1. Aeroelastic Model

An airfoil able to move in pitching and heaving degrees of freedom is placed on a pivot point passing through the pitching axis (z) (Figure 1). The equations of motion [5] for such an airfoil can be given as (Equations (1) and (2)):

$$F = m_h\ddot{y} + D_h\dot{y} + k_h y + S(\dot{\theta}^2 \sin\theta - \ddot{\theta}\cos\theta) \quad (1)$$

$$M = I_\theta\ddot{\theta} + D_\theta\dot{\theta} + k_\theta\theta - S\ddot{y}\cos\theta \quad (2)$$

where m_h is the heaving mass (kg), D is the damping (kg s^{-1} and kg m^2 s^{-1} rad^{-1} for pitching and heaving, respectively) and k is the stiffness coefficient (Nm^{-1} and Nm.rad^{-1} for pitching and heaving, respectively). I_θ (kg m^2) is the moment of inertia around the angular axis. The subscripts h and θ denote heaving and pitching. Plugging $S = m_p x_\theta$ (where m_p is the pitching mass (kg) and x_θ the centre of gravity location (m)), an inertial coupling exists in both degrees of freedom.

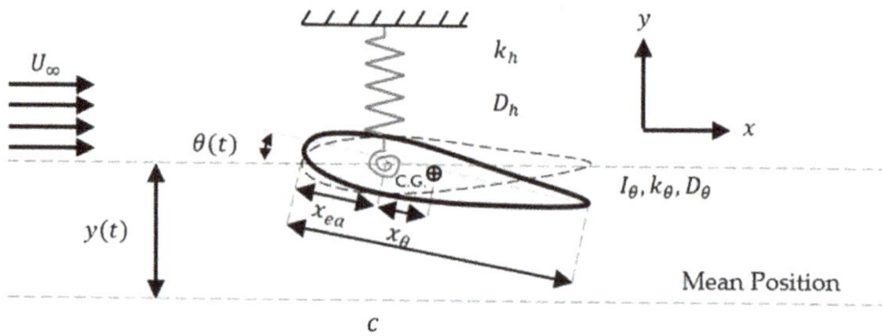

Figure 1. Basic representation of an inflexible, elastically mounted, symmetrical airfoil through figurative illustration of main parameters.

2.2. Oncoming Gust Shape

The gust shape presented in this work comprises two components [5], i.e., vertical and horizontal. The horizontal and vertical components (Equations (3) and (4)) were modelled by using the symmetric Gaussian distribution function:

$$u = a \times e^{[-(\frac{x-d}{n})^2]} \quad (3)$$

$$w = -a_0 \times e^{[-(\frac{x-d_0}{n_0})^2]} + a_1 \times e^{[-(\frac{x-d_1}{n_1})^2]} \quad (4)$$

where d (m) denotes the centroid position, a (m) denotes the amplitude and n (sec) denotes the time throughout which the components have a value above 50% of their peak amplitude, i.e., full duration at half maximum.

3. Numerical Procedure

The flow solution was performed on a numerical set of 143,416 structured cells consisting of NACA 0015 airfoil with a chord length of 0.12 m. The domain size was equal to 160c.

For the URANS solution, the pressure-based solver was utilised with the SIMPLE algorithm. The gust model was overlapped on the inlet velocity up till the critical flutter velocity. Numerical simulations for the velocity range of $(0.85 \le U/U_c \le 1)$ near the critical flutter velocity were performed (where U_c is critical velocity).

Acoustic data were acquired for 20,000 time steps along with a time step size duration of 10^{-5} s after the URANS model achieved steady state. The pressure was noted by the placement of acoustic receivers as described in the literature [6].

Model Validation

In order to validate the model, the coefficients of lift, drag and sound pressure levels were compared (Table 1) with those stated by experimental [7] and numerical setups [8,9]. Lastly, the sound pressure level in the one-third octave bands ($SPL_{1/3}$) were established and further related with the validation cases along with the mesh convergence study (Figure 2). The current model replicated the position of the main tone (~1.6 kHz) and sound pressure level (75 dB), which was in perfect harmony with the published data [7,10,11].

Table 1. Comparison of time averaged aerodynamic coefficients.

Coefficient	URANS (Existing Study)	Experimental [7]	% Difference	Numerical [8]	Numerical [9]
C_L	0.43	0.44	2.32%	0.45	0.46
C_D	0.0084	0.0083	1.20%	0.0075	0.0071

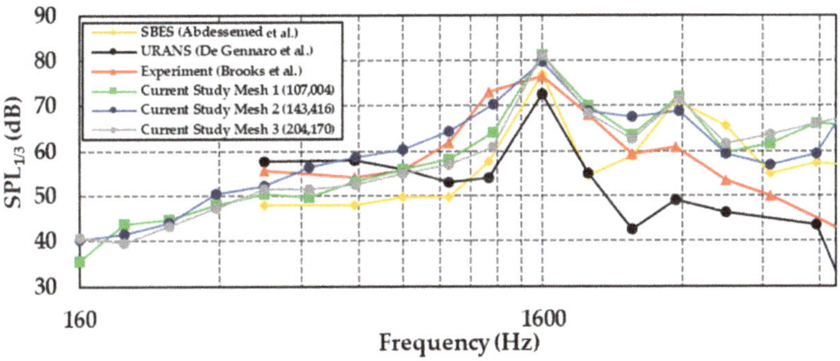

Figure 2. One-third octave band SPL showing SBES, URANS and experimental comparison for M2 along with mesh convergence study.

4. Results and Discussion

A fast Fourier transform (FFT) method was employed to determine the frequencies from the acoustic pressure signals. Here, the role of velocity (U/U_C), and the excitation frequency was investigated for the production of aerodynamic noise.

Production of Sound Waves

The current numerical simulations were performed for $Re = 80,000–120,000$ while varying U/U_c from 0.85 to 1. Figure 3 depicts a magnified schematic at $U = 0.85U_c$ for one of the 38 receivers at the circle (x = 5c). The forcing frequency (10.54 Hz) and its even harmonic (21.08 Hz) has a significant role in the production of the flow noise as the oncoming gust influences these signals for the lift and drag. This tendency is palpably seen for all receivers.

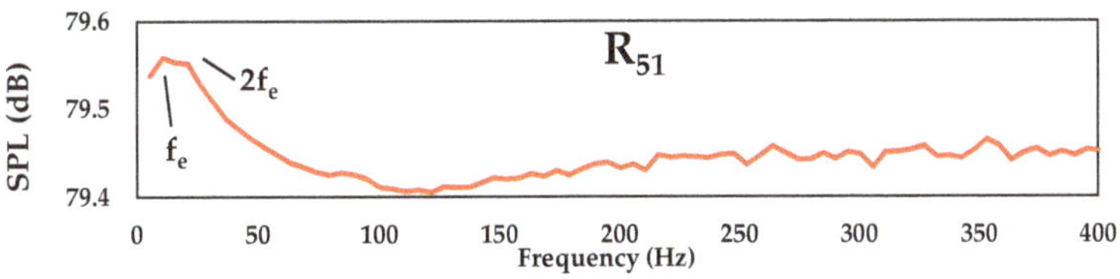

Figure 3. Frequency configuration of SPL about the flapping hydrofoil at one of the receiver locations in the middle circle.

An evaluation for the SPL for microphones positioned at the circles (x = 5c, 8c) is shown in Figure 4. The setting where the incoming velocity is equal to the critical velocity represents the highest perturbations in the flow media in terms of frequency and oscillation amplitude. With a growing (U/U_C), the SPL increases for the excitation frequencies. While traveling away from the airfoil, the amplitudes of SPL decrease.

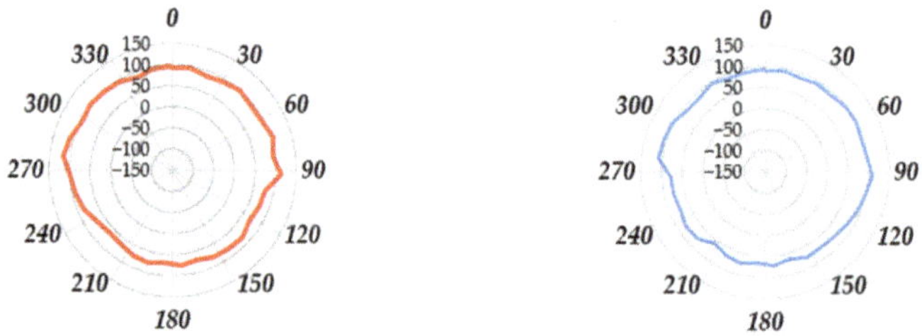

Figure 4. Polar plots showing the spread of SPL: the row expresses data for $\frac{U}{U_c} = 1$; columns 1 and 2 characterize measurements at the middle (x = 5c) and outer circles (x = 8c), respectively.

5. Conclusions

In this research work, the acoustic response of a passively moving airfoil subjected to a specific gust profile for a range of flow velocities was investigated numerically. The dependency of the sound pressure level magnitude on various parameters was explored. The flow velocity was one of the significant factors to control the change in the sound pressure level magnitudes. For the whole range of velocities (U/U_C), the sound pressure levels depicted the tonal noise on the excitation frequencies. The oscillation frequencies were a function of the force coefficients, which were basically reliant on the gust shape. The oscillation frequencies governed the whole spectrum of sound pressure levels. This aspect was the crux of this study and can be the pioneer phase to control the sound levels at will in real-world applications.

Author Contributions: Conceptualization, M.A. and K.A.; methodology, M.A. and S.M.; software, M.A. and K.A.; validation, M.A. and K.A.; formal analysis, M.E. and M.A.; investigation, M.A. and M.E.; data curation, M.A.; writing—original draft preparation, S.M.; writing—review and editing, M.A. and M.E.; supervision, M.E.; project administration, S.M. All authors have read and agreed to the published version of the manuscript.

Funding: This research received no external funding.

Institutional Review Board Statement: Not applicable.

Informed Consent Statement: Not applicable.

Data Availability Statement: Not applicable.

Conflicts of Interest: The authors declare no conflict of interest.

References

1. Kallas, S.; Geoghegan-Quinn, M. *Flightpath 2050 Europe's Vision for Aviation: Report of the High-Level Group on Aviation Research*; Publications Office, European Commission: Brussels, Belgium, 2011.
2. Métivier, V.; Dumas, G.; Poirel, D. Aeroelastic Dynamics of a NACA 0012 Airfoil at Transitional Reynolds Numbers. In Proceedings of the 39th AIAA Fluid Dynamics Conference, San Antonio, TX, USA, 22–25 June 2009.
3. Poirel, D.; Harris, Y.; Benaissa, A. Self-Sustained Aeroelastic Oscillations of a NACA0012 Airfoil at Low-to-Moderate Reynolds Numbers. *J. Fluids Struct.* **2008**, *24*, 700–719. [CrossRef]
4. Poirel, D.; Métivier, V.; Dumas, G. Computational Aeroelastic Simulations of Self-Sustained Pitch Oscillations of a NACA0012 at Transitional Reynolds Numbers. *J. Fluids Struct.* **2011**, *27*, 1262–1277. [CrossRef]
5. Manzoor, S.; Sheikh, N.; Ali, M. Numerical Investigation of Transient Response of a Coupled Two-Degrees-of Freedom Symmetric Airfoil before Flutter. *Int. J. Aeroacoust.* **2018**, *17*, 275–294. [CrossRef]
6. Khalid, M.S.U.; Akhtar, I.; Wu, B. Quantification of Flow Noise Produced by an Oscillating Hydrofoil. *Ocean. Eng.* **2019**, *171*, 377–390. [CrossRef]

7. Brooks, T.; Pope, D.; Marcolini, M. Airfoil Self-Noise and Prediction. Available online: https://ntrs.nasa.gov/citations/19890016302 (accessed on 15 August 2022).
8. Abdessemed, C.; Bouferrouk, A.; Yao, Y. Aerodynamic and Aeroacoustic Analysis of a Harmonically Morphing Airfoil Using Dynamic Meshing. *Acoustics* **2021**, *3*, 177–199. [CrossRef]
9. de Gennaro, M.; Kuehnelt, H.; Zanon, A. Numerical Prediction of the Tonal Airborne Noise for a NACA 0012 Aerofoil at Moderate Reynolds Number Using a Transitional URANS Approach. *Arch. Acoust.* **2017**, *42*, 653–675. [CrossRef]
10. Arbey, H.; Bataille, J. Noise Generated by Airfoil Profiles Placed in a Uniform Laminar Flow. *J. Fluid Mech.* **1983**, *134*, 33–47. [CrossRef]
11. Paterson, R.W.; Vogt, P.G.; Fink, M.R.; Munch, C.L. Vortex noise of isolated airfoils. *J. Aircr.* **1973**, *10*, 296–302. [CrossRef]

engineering proceedings

Proceeding Paper

Design and Development of a Control Volume Spray Chamber (CVSC) for Fuel Spray Visualization †

Muteeb Ul Haq, Ali Turab Jafry *, Saad Ahmad and Atif Muzaffar

Faculty of Mechanical Engineering, Ghulam Ishaq Khan Institute of Engineering Sciences and Technology, Topi 23640, Pakistan
* Correspondence: ali.turab@giki.edu.pk
† Presented at the 2nd International Conference on Advances in Mechanical Engineering (ICAME-22), Islamabad, Pakistan, 25 August 2022.

Abstract: Biofuels have attracted significant attention in recent years as a potential replacement of fossil fuels. In order to test a new fuel, it is necessary to know its physiochemical properties. Fuel spray atomization of any fuel is the most critical process that has a direct effect on the fuel–air mixing ratio, combustion and emissions. This research presents the design and development of a control volume spray chamber (CVSC) for analyzing macroscopic fuel spray characteristics under varying operating conditions. The spray results reveal that the fuel penetration length (FPL) and spray cone angle (SCA) both decreased with the increase in ambient pressure, but when the injection pressure was increased, a longer FPL and wider SCA were observed. The light intensity levels revealed broader and higher drop densities at the axial distance of 40 mm from the nozzle.

Keywords: control volume spray chamber; spray characteristics; fuel penetration length; spray cone angle

check for updates

Citation: Haq, M.U.; Jafry, A.T.; Ahmad, S.; Muzaffar, A. Design and Development of a Control Volume Spray Chamber (CVSC) for Fuel Spray Visualization. *Eng. Proc.* **2022**, *23*, 35. https://doi.org/10.3390/engproc2022023035

Academic Editors: Mahabat Khan, M. Javed Hyder, Muhammad Irfan and Manzar Masud

Published: 27 September 2022

1. Introduction

Nowadays, biofuels attract many engine manufacturers as an alternative fuel to diesel due to their biodegradable nature. They are clean and renewable source of energy. Biodiesel comprises an excess of oxygen that enhances combustion and reduces the particulate matter and soot emissions [1]. Biofuels usually have a higher fuel viscosity and surface tension which affects the fuel injection process and spray quality when used in a diesel engine.

Spray characteristics have a major effect on the combustion process which defines the thermal efficiency of an engine. In order to use biofuels or their blends in a diesel engine, it is necessary to study the fuel spray characteristics that will provide a better insight into their performance and emission properties. Wang et al. [2] developed a control volume vessel (CVV) to study the spray behavior of palm oil and cooked oil and found that a greater cone angle and a larger drop size are observed for biodiesels when compared to diesel. Agarwal et al. [3] studied the spray behavior of Karanja biodiesel in a constant volume spray visualization chamber comprising four optical windows and using a multi-hole injection nozzle. Lee et al. [4] investigated the macroscopic spray properties of biodiesel and its blends obtained from soybean and canola oil in a pressurized spray chamber with five optical windows. FPL is the distance covered from the tip of the nozzle to 95% of the farthest distance the fuel has reached, while SCA is the angle between the two lines starting from the nozzle tip and passing through the periphery of the spray until the half point of the penetration length.

The effect of viscosity and surface tension on the fuel spray properties of castor oil, neem oil and sunflower oil was revealed by Das et al. [5]. Bohl et al. [6] used a control volume vessel with four optical windows of 100 mm diameter to study the spray behavior of palm oil, used cooking oil and soyabean. Biofuels that are being produced from edible sources are less likely to be used on a commercial scale; that is why biofuels derived from nonedible sources are more popular and beneficial. In this study, a control volume spray vessel (CVSV) with three optical windows was developed to analyze the macroscopic spray properties of diesel.

2. Materials and Methods

First, a CAD model of the CVSC was made using SolidWorks (2021, Dassault Systèmes, Vélizy-Villacoublay, France). An exploded view of CVSC assembly can be observed in Figure 1.

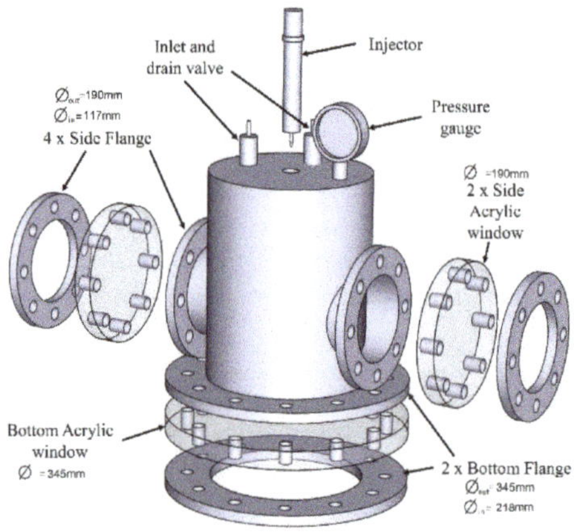

Figure 1. Schematic diagram of CVSC with fabrication details.

2.1. Fabrication of Experimental Setup

CVSC was fabricated at our workshop in GIK Institute. Several processes involved during fabrication can be observed in a Figure 2. The process started with a hollow cylinder made of cast iron shown in Figure 2b. Two holes were made in the cylinder using gas welding for the optical windows, after which 2 small cylinders were welded to these windows as shown in Figure 2d. Then, the side flanges and the bottom flange were welded, which can be observed in Figure 2e,f. Figure 2g,h shows the cutting of the acrylic window on the lathe and drill machine. Finally, the top plate of the chamber was cut with a hole in the center for the injector, as shown in Figure 2i,j.

Figure 2. Design materials and processes involved during fabrication of CVSC. (**a**) 4 small and 2 large flanges of CVSC windows, (**b**) Cylindrical body of CVSC, (**c**) Holes being made on cylindrical body for side windows, (**d**) Grinding from the inner side of CVSC after welding, (**e**) Removal of slag after welding of 2 side flange, (**f**) 3 flanges welded to the CVSC, (**g**,**h**) Cutting of acrylic sheet for the CVSC windows, (**i**) Drilling of hole on the top plate for injector mounting, (**j**) Injector mounted on the top of CVSC.

Figure 3a,b shows the front view and side view of the CVSC, respectively. The inside and the outside views of CVSC from the bottom are shown in Figure 3c,d, respectively.

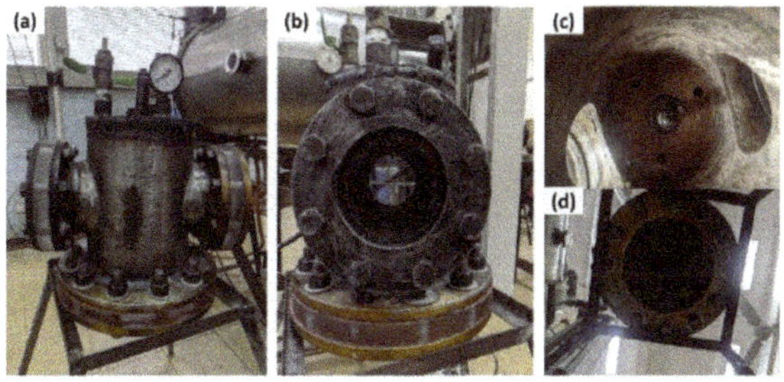

Figure 3. Various orientations of CVSC. (**a**) Sealed CVSC rested on a stand, with pressure gauge, injector and drain valves on the top, (**b**) Side view of CVSC, (**c**) Inside view of CVSC from the bottom window and (**d**) Bottom view of CVSC.

2.2. Experimental Setup

The CVSC was mounted on the stand, as shown in Figure 3a. Two 100 W LED lights were used for illuminating the chamber through side windows. An injector with 0.29 mm hole diameter was used. Video recording of the spray process was captured at 960 frames per second through the bottom side of CVSC. Images were extracted from the slow-motion videos using DaVinci Resolve software (Blackmagic Design, 18.0.1v, Victoria, Australia). Spray images were then processed using Image J software (National Institutes of Health and LOCI, 1.48 v, Bethesda, MD, USA, and Madison, WI, USA) to quantify the macroscopic spray properties of the jet.

3. Results

Fuel Spray Analysis of Diesel

Diesel is tested in the CVSC to study the macroscopic spray behavior, the injection and ambient conditions along with the fuel properties, which are given in Table 1. It can be seen from Figure 4a,c that a higher injection pressure increases the FPL because, at higher pressures, the fuel is injected with a greater force and possesses a higher momentum, allowing it to penetrate further. When the ambient pressure is increased, incoming fuel jet faces more drag, due to which the jet expands radially and the FPL decreases. The injection pressure also increases the SCA; however, the ambient pressure has a dominant effect. Figure 4b shows the spray density in radial direction in terms of light intensity levels at axial locations of 10 mm and 40 mm. The light intensity closer to the nozzle is narrow because it did not have enough time to expand, while at 40 mm, a broader light intensity can be seen for both the ambient pressures. A maximum intensity was observed at an ambient pressure of 8 bar and at the axial location of 40 mm, while the lowest was recorded for a 2-bar ambient pressure at the axial location of 10 mm.

Table 1. The injection and ambient conditions along with the fuel properties for analyzing the fuel spray in a CVSC.

Parameter	Quantity	Parameter	Quantity
Injection pressure	500, 1000 bar	Injection duration	2 ms
Injection temperature	300 K	Chamber diameter	218 mm
Ambient pressure	2 & 8 bar	Nozzle diameter	0.290 mm
Chamber temperature	300 K	Diesel density	837.9 kg/cm^3

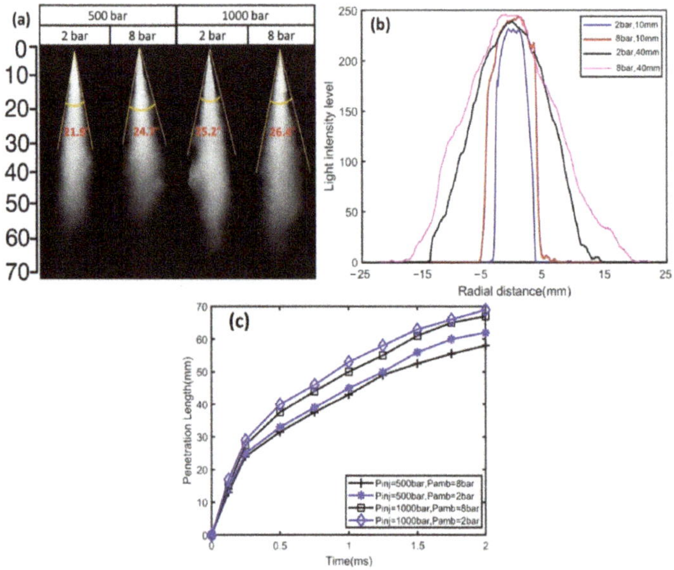

Figure 4. (**a**) Experimental images of diesel spray showing PL and SCA for two different injection and ambient pressures. (**b**) Light intensity levels with regard to radial distance for 2 axial locations and ambient pressures for 1000 bar injection pressure. (**c**) PL for diesel spray against varying injection and ambient pressures.

4. Conclusions

The three-window design of the CVSC was effective for fuel spray visualization for both single and multi-hole injectors. An increase in injection pressure enhanced the FPL and SCA while an increase in ambient pressure caused a reduction in the FPL and SCA. The light intensity, which depicts the spray density, was narrow when closer to the nozzle and expanded in the axial direction. The maximum intensity is at the core of the spray and decreases in the radial direction, as the spray is less dense at the outer periphery.

Author Contributions: Conceptualization, M.U.H. and A.T.J.; methodology, M.U.H. and A.M.; software, M.U.H. and S.A.; validation, M.U.H. and S.A.; formal analysis, M.U.H. and A.T.J.; investigation, M.U.H. and A.M.; data curation, A.T.J.; writing—original draft preparation, M.U.H.; writing—review and editing, M.U.H. and A.T.J.; supervision, A.T.J. All authors have read and agreed to the published version of the manuscript.

Funding: This research received no external funding.

Institutional Review Board Statement: Not applicable.

Informed Consent Statement: Not applicable.

Data Availability Statement: Not applicable.

Conflicts of Interest: The authors declare no conflict of interest.

References

1. McCormick, R.L.; Graboski, M.S.; Alleman, T.L.; Herring, A.M.; Tyson, K.S. Impact of biodiesel source material and chemical structure on emissions of criteria pollutants from a heavy-duty engine. *Environ. Sci. Technol.* **2001**, *359*, 1742–1747. [CrossRef] [PubMed]
2. Wang, X.; Huang, Z.; Kuti, O.A.; Zhang, W.; Nishida, K. Experimental and analytical study on biodiesel and diesel spray characteristics under ultra-high injection pressure. *Int. J. Heat Fluid Flow* **2010**, *31*, 659–666. [CrossRef]

3. Agarwal, A.K.; Dhar, A.; Gupta, J.G.; Kim, W.I.; Lee, C.S.; Park, S. Effect of fuel injection pressure and injection timing on spray characteristics and particulate size–number distribution in a biodiesel fuelled common rail direct injection diesel engine. *Appl. Energy* **2014**, *130*, 212–221. [CrossRef]
4. Lee, D.; Jho, Y.; Lee, C.S. Effects of soybean and canola oil–based biodiesel blends on spray, combustion, and emission characteristics in a diesel engine. *J. Energy Eng.* **2014**, *140*, A4014012. [CrossRef]
5. Das, M.; Sarkar, M.; Datta, A.; Santra, A.K. Study on viscosity and surface tension properties of biodiesel-diesel blends and their effects on spray parameters for CI engines. *Fuel* **2018**, *220*, 769–779. [CrossRef]
6. Bohl, T.; Tian, G.; Smallbone, A.; Roskilly, A.P. Macroscopic spray characteristics of next-generation bio-derived diesel fuels in comparison to mineral diesel. *Appl. Energy* **2017**, *186*, 562–573. [CrossRef]

engineering proceedings

MDPI

Proceeding Paper

Experimental Investigation of Direct Contact Condensation Using a Square Steam Nozzle †

Noman Arif Khan [1,*], Ajmal Shah [2], Abdul Quddus [1], Haseeb Afzal [1], Shumail Hassan [1], Muhammad Khawar Ayub [3] and Mazhar Iqbal [1]

1 Department of Mechanical Engineering, Pakistan Institute of Engineering and Applied Sciences (PIEAS), Islamabad 45650, Pakistan
2 Center for Mathematical Sciences, Pakistan Institute of Engineering and Applied Sciences (PIEAS), Islamabad 45650, Pakistan
3 Theoretical Physics Division, Pakistan Institute of Nuclear Science & Technology (PINSTECH), Islamabad 45650, Pakistan
* Correspondence: nomanarifkhan98@gmail.com

† Presented at the 2nd International Conference on Advances in Mechanical Engineering (ICAME-22), Islamabad, Pakistan, 25 August 2022.

Abstract: Direct-contact condensation (DCC) has acquired an important role in the industrial sector due to its high mass and heat transfer rates. In this paper, the influence of steam pressure and water temperature on cavity shapes were studied from symmetrical and diagonal plane views. The cavity shapes observed were oscillatory, conical, ellipsoidal, and double expansion–contraction. The recompression shock wave at nozzle corners was found to cause steam cavity compression in the diagonal plane. The dimensionless penetration length was found to increase with the rise in steam pressure and water temperature and lay in the range from 3.38 to 5.55. The experimental data of dimensionless penetration length was in good agreement with previous correlations.

Keywords: supersonic nozzle; direct-contact condensation; recompression shock wave; intercepting shock wave; condensation potential; cavity penetration length; cavity plume shape

Citation: Khan, N.A.; Shah, A.; Quddus, A.; Afzal, H.; Hassan, S.; Ayub, M.K.; Iqbal, M. Experimental Investigation of Direct Contact Condensation Using a Square Steam Nozzle. *Eng. Proc.* **2022**, *23*, 29. https://doi.org/10.3390/engproc2022023029

Academic Editors: Mahabat Khan, M. Javed Hyder, Muhammad Irfan and Manzar Masud

Published: 22 September 2022

1. Introduction

Steam water direct-contact condensation (DCC) is a thermal hydraulic phenomenon that occurs when saturated/superheated steam is injected into subcooled water. Kerney et al. [1] conducted a pioneering study on DCC and presented a correlation for cavity penetration length. Kim et al. [2] presented empirical correlations for cavity penetration length and the average heat transfer coefficient for sonic nozzles.

Wu et al. [3] conducted an extensive study for supersonic nozzles and showed that cavity shapes were dependent upon shock and expansion waves at the nozzle exit. Quddus et al. [4] discussed the effect of the nozzle angle on DCC using a bevelled steam nozzle. Xu et al. [5] measured the heat transfer coefficient and penetration length using numerical investigation. Tsutsumi et al. [6] studied a square nozzle from both an experimental and Computational Fluid Dynamics (CFD) approach and obtained the shock structures on a diagonal and symmetrical plane.

In this current study, a supersonic square nozzle was used for steam injection due to its enhanced mixing and entrainment [7]. The influence of steam pressure and water temperature on cavity shapes and cavity penetration length were studied using image capturing and processing. The results of the current experimental study will help in the better designing of DCC-based industrial components with safer operation.

2. Materials and Methods

The experimental setup, shown in Figure 1, was designed to provide saturated steam injection in subcooled water. The electric boiler could provide 52 kg/h of steam (~99% quality), at a maximum pressure of 8 bar. The electric boiler was a cylindrical tank containing four electric heaters (9 kW capacity of one heater) submerged in water. Steam cavity was captured using a high-speed camera and processed using a MATLAB code. The operating conditions and nozzle dimensions are given in Table 1.

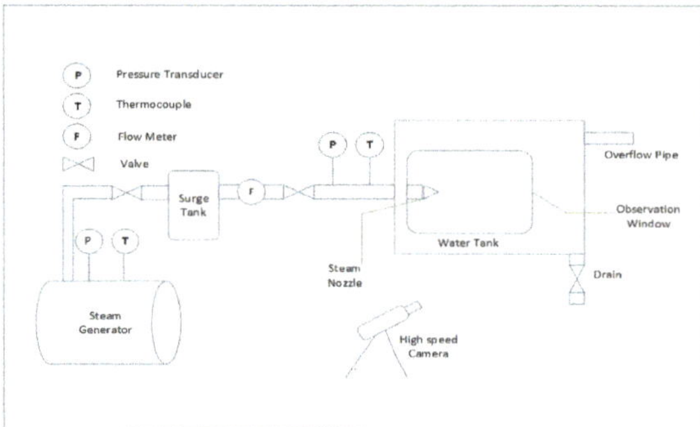

Figure 1. Experimental Setup for DCC.

Table 1. Operating conditions and nozzle dimensions.

Parameters	Value/Range
Steam pressure (Absolute)	1.5–4.5 bar
Water temperature, T_W	35 °C and 55 °C
Nozzle inlet dimensions	10 mm × 12 mm
Nozzle throat dimensions	5 mm × 5 mm
Nozzle exit dimensions	5.25 mm × 5.25 mm

3. Results and Discussion

In this section, the influence of steam pressure and water temperature on the cavity shapes and penetration length is discussed. A steam cavity was observed from the symmetrical plane and diagonal plane view [6]. Buoyancy effects were negligible at T_W = 35 °C and 55 °C.

3.1. Influence of Steam Pressure and Water Temprature on the Cavity Shapes

As shown in Figure 2, the symmetrical plane view was captured at T_W = 35 °C. At 1.5 bar, as shown in Figure 2a, oscillatory condensation occurred. At 2.5 bar, as shown in Figure 2b, a conical shape was observed, due to high degree of subcooling of water. At 3.5 bar, as shown in Figure 2c, an ellipsoidal shape was observed. The nozzle exit pressure was higher than ambient water pressure, so an expansion wave formed at the edges of the nozzle and steam expanded. Expansion waves interacted with the cavity boundary to make an intercepting shock wave and steam contracted. At 4.5 bar, as shown in Figure 2d, an ellipsoidal shape was observed with high expansion due to the stronger expansion waves.

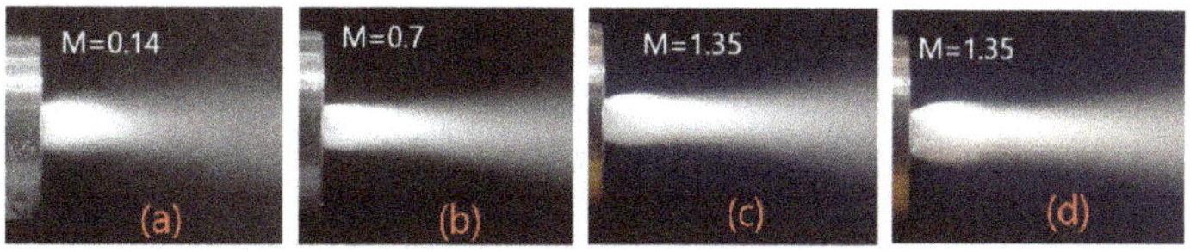

Figure 2. Symmetrical plane view. Steam pressure (bar) effect at T_W = 35 °C (**a**) 1.5 (**b**) 2.5 (**c**) 3.5 (**d**) 4.5.

As shown in Figure 3, a diagonal plane view was captured at T_W = 35 °C. At the nozzle corners, recompression shock waves formed due to the interaction of the expansion waves of the two adjacent edges, which formed an overexpanded region [6]. At 2.5 bar, as shown in Figure 3b, the cavity was conical due to recompression shock waves. At 3.5 bar, as shown in Figure 3c, the cavity was conical, but the cavity from the symmetrical plane view was ellipsoidal. This was due to recompression shock waves at the nozzle corners. In the diagonal plane, an intercepting shock wave also formed after the interaction of expansion waves with cavity boundary. The recompression shock wave at the nozzle corner as well as the intercepting shock wave contracted the steam. At 4.5 bar, as shown in Figure 3d, the cavity shape observed was ellipsoidal, but it was actually conical. At high steam pressure, expansion waves are stronger, resulting in stronger recompression shock waves at the corners. The expansion from the edges coming in front of diagonal plane is shown in Figure 3d. At 3.5 bar, as shown in Figure 3c, the expansion was small and did not appear in the diagonal plane view.

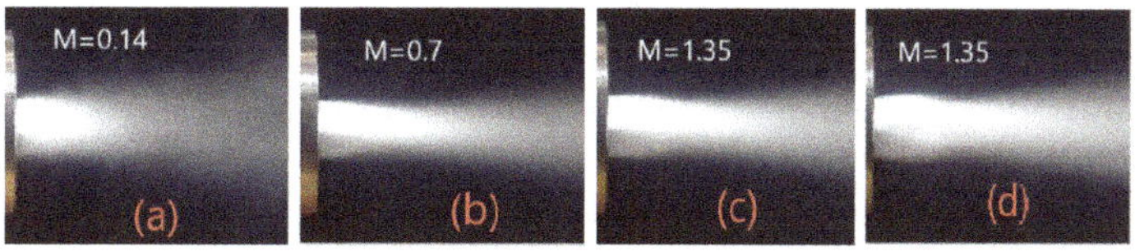

Figure 3. Diagonal plane view. Steam pressure (bar) effect at T_W = 35 °C (**a**) 1.5 (**b**) 2.5 (**c**) 3.5 (**d**) 4.5.

As shown in Figure 4, the symmetry plane view was captured at T_W = 55 °C. At 1.5 bar, as shown in Figure 4a, condensation oscillation was found to be more violent due to the lower condensation potential at a higher water temperature. At 2.5 bar, as shown in Figure 4b, the cavity was ellipsoidal. This is due to the decrease in condensation potential, which increases the interface surface area for dissipating the heat coming from the steam. The increase in interface surface area was achieved by increasing the expansion and penetration length. At 3.5 bar, as shown in Figure 4c, the cavity is found to be a double expansion–contraction due to the addition of momentum and heat at high steam pressure. The cavity first expanded to cater for the extra heat and then it was compressed by the ambient water pressure. It then expanded again, as the pressure recovery was higher at high water temperature. At 4.5 bar, as shown in Figure 4d, the cavity shape was again double expansion–contraction, but with a higher expansion to cater for the addition of more heat.

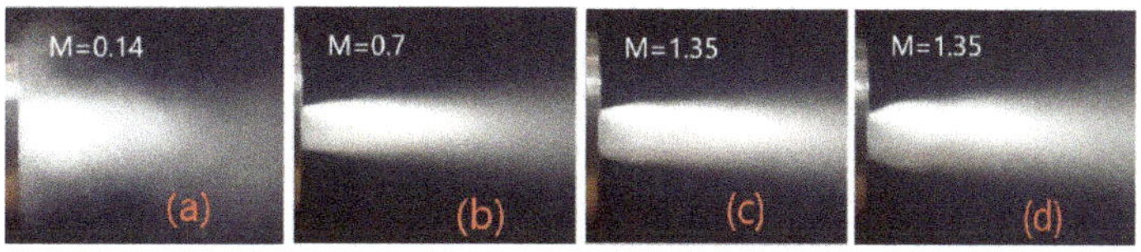

Figure 4. Symmetry plane view. Steam pressure (bar) effect at T_W = 55 °C (**a**) 1.5 (**b**) 2.5 (**c**) 3.5 (**d**) 4.5.

As shown in Figure 5, the diagonal plane view was captured at T_W = 55 ° C. At 2.5 bar, as shown in Figure 5b, the cavity was conical but ellipsoidal from the symmetry plane view, due to the recompression shock wave at the corners. At 3.5 bar and 4.5 bar, as shown in Figures 5c and 5d, expansion from the edges came in front of the diagonal plane.

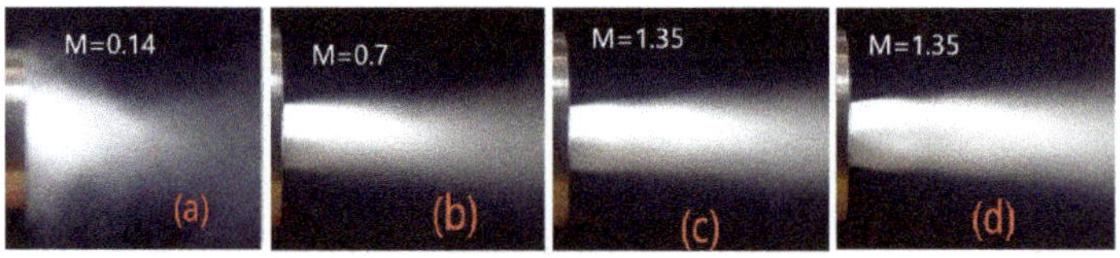

Figure 5. Diagonal plane view. Steam pressure (bar) effect at T_W = 55 °C (**a**) 1.5 (**b**) 2.5 (**c**) 3.5 (**d**) 4.5.

3.2. *Influence of Steam Pressure and Water Temperature on the Penetration Length*

The variation in the dimensionless penetration length is shown in a black line in Figure 6. The dimensionless penetration length was obtained by dividing it by the width of nozzle. As the steam pressure increased, the penetration length increased due to the increase in momentum transfer. At low temperatures, the penetration length was small due to the high condensation potential. At high temperatures, the interface area increased, which lead to a large degree of penetration. The dimensionless penetration length was found to be in the range of 3.38–5.55. The data lies in the range from −8.87% to +20.3% range with absolute deviation of 13.1%, when compared with correlation of Kerney and Kim.

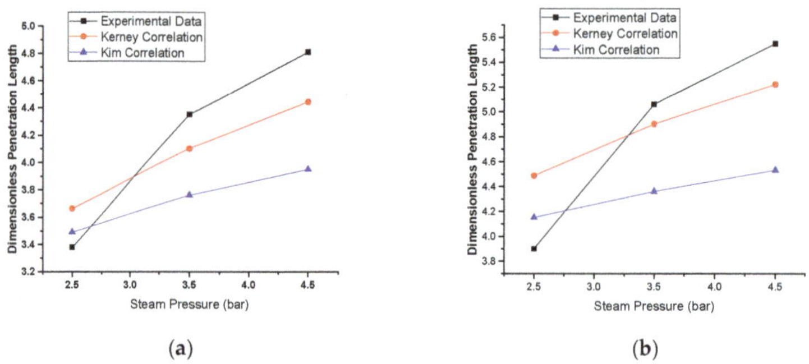

Figure 6. Variation of current data with predicted data (**a**) T_W = 35 °C (**b**) T_W = 55 °C.

4. Conclusions

Four cavity shapes were observed—oscillatory, conical, ellipsoidal, and double expansion–contraction. The steam cavity from the diagonal plane view was conical for all operating conditions due to recompression wave at the corner of the nozzle exit. The expansion captured in the diagonal plane view was that of the expansion from the nozzle edges at higher steam pressures. The penetration length increased with the rise in steam pressure and water temperature.

Author Contributions: N.A.K.: Conceptualization, Data curation, Methodology, Investigation, Formal analysis, Writing—original draft. A.S.: Investigation, Funding acquisition, Supervision, Conceptualization. A.Q.: Conceptualization, Project administration, Supervision. H.A.: Data curation. S.H.: Funding acquisition, Resources, Review. M.K.A.: Data curation, Writing and Review Editing. M.I.: Resources. All authors have read and agreed to the published version of the manuscript.

Funding: This research received no external funding.

Conflicts of Interest: The authors declare no conflict of interest.

References

1. Kerney, P.; Faeth, G.; Oslon, D. Penetration characteristics of a submerged steam jet. *AIChE J.* **1972**, *18*, 548–553. [CrossRef]
2. Kim, H.Y.; Bae, Y.Y.; Song, C.H.; Park, J.K.; Choi, S.M. Experimental study on stable steam condensation in a quenching tank. *Int. J. Energy Res.* **2001**, *25*, 239–252. [CrossRef]
3. Wu, X.Z.; Yan, J.J.; Shao, S.F.; Cao, Y.; Liu, J.P. Experimental study on the condensation of supersonic steam jet submerged in quiescent subcooled water: Steam plume shape and heat transfer. *Int. J. Multiph. Flow* **2007**, *33*, 1296–1307. [CrossRef]
4. Quddus, A.; Shah, A.; Qureshi, K.R.; Ayub, M.K.; Tahir, A.; Iqbal, M. Study of steam jet characteristics and regime maps for bevelled spray nozzles exhausting into quiescent water. *Int. J. Heat Mass Transf.* **2022**, *190*, 122780. [CrossRef]
5. Xu, Q.; Liang, L.; She, Y.; Xie, X.; Guo, L. Numerical investigation on thermal hydraulic characteristics of steam jet condensation in subcooled water flow in pipes. *Int. J. Heat Mass Transf.* **2022**, *184*, 122277. [CrossRef]
6. Tsutsumi, S.; Teramoto, S.; Yamaguchi, K.; Nagashima, T. Structure of underexpanded jets from square nozzles. *AIAA J.* **2006**, *44*, 1287–1291. [CrossRef]
7. Zhang, H.; Chen, Z.; Jiang, X.; Guo, Z. The initial flow characteristics of supersonic jets with different geometries. *Phys. Lett. A* **2015**, *379*, 1256–1262. [CrossRef]

engineering proceedings

MDPI

Proceeding Paper

Experimental Study of Steam–Water Direct Contact Condensation in a Horizontal Pipe Geometry †

Shumail Hassan [1,*], Mazhar Iqbal [1], Ajmal Shah [2], Abdul Quddus [1], Noman Arif Khan [1], Haseeb Afzal [1] and Muhammad Khawar Ayub [3]

1 Department of Mechanical Engineering, Pakistan Institute of Engineering and Applied Sciences (PIEAS), Islamabad 45650, Pakistan
2 Center for Mathematical Sciences, Pakistan Institute of Engineering and Applied Sciences (PIEAS), Islamabad 45650, Pakistan
3 Theoretical Physics Division, Pakistan Institute of Nuclear Science & Technology (PINSTECH), Islamabad 45650, Pakistan
* Correspondence: shumailhassan9@gmail.com

† Presented at the 2nd International Conference on Advances in Mechanical Engineering (ICAME-22), Islamabad, Pakistan, 25 August 2022.

Abstract: The phenomenon of direct contact condensation (DCC) has an advantageous feature of high heat, mass, and momentum transfer efficiencies and hence is highly significant for various steam-related industries such as chemical and nuclear industries. The present work investigates the underlying physics of steam plume shapes during steam–water DCC of saturated steam injection into a subcooled water-filled restricted geometry. These experiments have been performed using an orifice-type nozzle for saturated steam injection into a circular, horizontal pipe. To study the effects of pressure and the degree of subcooling of water on the steam plumes, the performed study utilized initial steam pressure and water temperature in the ranges of 1–2 bars and 60–70 °C, respectively. Numerous plume shapes such as conical, elliptical, and divergent are observed under different experimental conditions, which elongate and extend at higher subcooling temperatures. The temperature distribution within the test section as a result of steam injection has also been studied. The condensation-induced water hammer (CIWH) has also been observed under various conditions in terms of a propagating pressure oscillation.

Keywords: multiphase flow; direct contact condensation; plume shape; thermal hydraulics

Citation: Hassan, S.; Iqbal, M.; Shah, A.; Quddus, A.; Khan, N.A.; Afzal, H.; Ayub, M.K. Experimental Study of Steam–Water Direct Contact Condensation in a Horizontal Pipe Geometry. *Eng. Proc.* **2022**, *23*, 25. https://doi.org/10.3390/engproc2022023025

Academic Editors: Mahabat Khan, M. Javed Hyder, Muhammad Irfan and Manzar Masud

Published: 21 September 2022

1. Introduction

DCC is a complex and significant two-phase phenomenon that finds applications in many industrial setups such as steam de-superheating systems, condensate recovery systems, and steam-jet pumps, and it is quite important in applications such as district heating systems (DHSs) for domestic heating requirements and in emergency core cooling systems (ECCSs) of nuclear power plants.

In the past few decades, researchers have studied different aspects of this subject and mainly worked on the assessment of steam plume geometry, evaluation of average DCC heat transfer coefficient, and estimation of condensation regime maps [1]. Many numerical [2,3] and experimental [4,5] studies have been performed, until recently, on steam injection into quiescent water 'pools'. In case of 'restricted geometries', the previous studies were based upon the injection of steam into water to demonstrate the different condensation regimes and modes [6]. Unlike past studies, Datta [7] attempted to study different parameters of subcooled water being injected into a horizontal pipe filled with steam, but he did not study the injection of steam into a geometry having subcooled water.

Several experimental facilities such as PMK-2 at the Hungarian Atomic Energy Institute [8,9] have been established to study the pressure transients during steam–water DCC, i.e., condensation-induced water hammers (CIWHs). A CIWH is a precursor to stratified-to-slug flow transition along with increased flow instability making the condensation physics more complicated.

In this study, a circular, horizontal pipe is selected for the investigation of DCC steam plume shapes, and steam is injected via a single-hole, orifice-type nozzle into the subcooled water. A deep understanding of steam plume shapes, temperature, and pressure distribution is crucial for technological advancement in the safe design and operation procedures of nuclear and other steam-related industries.

2. Materials and Methods

The experimental facility setup at PIEAS is shown schematically in Figure 1. The parameters shown in Table 1 are selected because they are the most commonly occurring conditions in the steam condensate recovery systems wherein steam is present at relatively higher temperatures (60–70 °C) and lower pressures (1–2 bar). The setup majorly comprises an electric boiler with a surge tank, a circular Perspex pipe (main test section), an orifice-type nozzle, temperature and pressure instrumentation, and steam medium piping. The transparent Perspex pipe (72 inch long and $2\frac{1}{4}$ inch OD; L/D~36) resembles Datta's [7] test section (stainless steel pipe having L/D~30) and has K-type thermocouples (M8, probe length = 1 inch, ~0.5 K accuracy) and four pressure transmitters (0.2% FS accuracy) installed on it; all instruments were duly calibrated, captured by high-speed camera, and recorded by a DAQ system (Pangu Automation System Co., Ltd., VX8124R, Hangzhou, China). The saturated steam was injected horizontally as a stable sonic jet in the Perspex pipe via a nozzle. Steam plume shapes were captured for horizontal steam injection against a water degree of subcooling at different steam injection pressures.

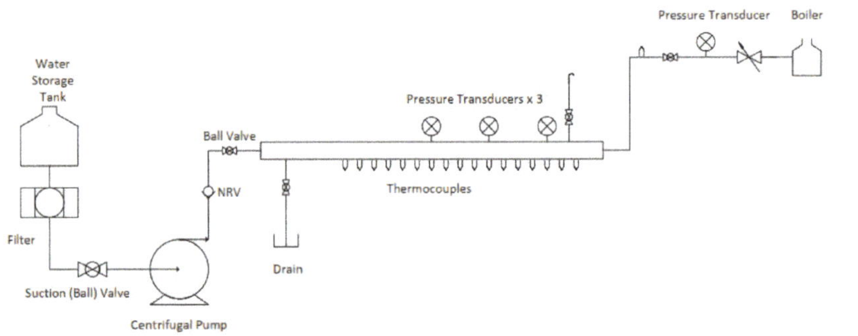

Figure 1. Schematic representation of DCC setup.

Table 1. Particulars of Operating Conditions.

Parameters	Values
Steam Injection Pressure	1–2 bar
Water Temperature (Degree of Subcooling)	60–70 °C
Inlet Diameter of Nozzle	$\frac{1}{2}$ inch
Exit Diameter of Nozzle	3 mm

3. Results and Discussion

Different steam plume shapes obtained are captured through a high-speed camera because the system is highly turbulent and changes the water subcooling degree in a matter of milliseconds. Elongated steam plumes are obtained for lesser degrees of subcooling, i.e., at elevated water temperatures, quite possibly due to decreased condensation capacity at higher temperatures. At lower to intermediate pressures, the plume becomes inclined at an angle and ultimately stabilizes itself at relatively higher pressure values and increased water temperatures. Three major types of steam plumes have been observed: conical, elliptical, and divergent, as shown in Figure 2. The formation of conical plumes can be attributed to higher degrees of subcooling or condensation potential at lower pressure values immediately at the start of steam injection. However, the conical plumes are converted into elliptical and divergent at increasing pressure values and water temperature as a consequence of decreasing condensation potential. There is an observable temperature distribution along the pipe length following the steam injection, as shown in Figure 2, which shows the readings of thermocouples installed along the length of pipe during steam injection, at the instant when steam plumes are captured; i.e., the ambient water temperature rises as steam is added, thereby changing the degree the subcooling of water. In other words, Figure 3 indicates steam-front propagation within the test section.

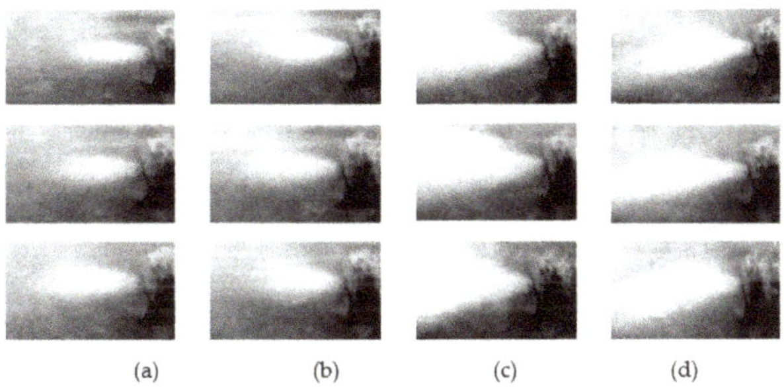

Figure 2. Variations in steam plumes with increasing water temperature (plume elongation) at 1.5 bar: (**a**) conical plumes; (**b**) elliptical plumes; (**c**) divergent plumes; (**d**) plumes with an inclination angle.

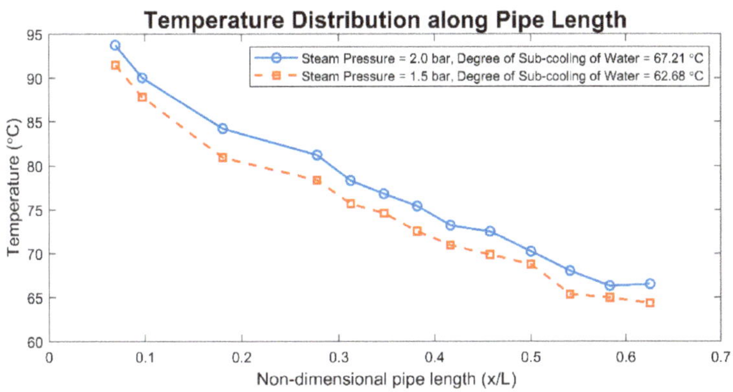

Figure 3. Temperature distribution along pipe length for two operating conditions.

In terms of validation of our setup and results, the temperature distribution obtained in the DCC-1 experiment (at 2 bar steam pressure and water temperature of 29.97 °C) of Datta et al. [7] is nearly similar to our data (at 2 bar steam pressure, water temperature of 32.79 °C). Similarly, no prior experimental work is available to validate the plume study for horizontal geometries. Considering the novelty of this field, some computational studies slightly confirm our estimated plume shapes: a recent work [10] utilized CFD simulation in a vertical pipe under stable sonic jet conditions and reached conclusions, i.e., elongation of plumes with injection pressure, transition of conical plumes with increasing pressures, etc., that conform to our findings.

4. Conclusions

In this experimental investigation, different steam plume shapes have been observed for saturated steam injected into subcooled water in a circular, horizontal geometry. The dependence of steam plume shapes on the steam injection pressure and degree of subcooling of water medium has been studied in a circular pipe. Some of the significant conclusions from the experimental study can be outlined as follows:

- Under these experimental conditions, three different steam plumes have been identified, i.e., conical, elliptical, and divergent.
- The temperature distributions obtained for two different steam pressure and water temperature values show decreasing trends along pipe length in a similar fashion.
- The steam ejected from the nozzle can be deflected downwards (or upwards) at an inclination from the reference nozzle axis due to high turbulence and CIWH occurring.
- Steam plumes elongate and extend as the temperature of subcooled water increases.

Author Contributions: S.H.: data curation, formal analysis, experimentation, investigation, writing—final draft. A.S.: conceptualization, methodology, supervision. A.Q.: conceptualization, funding acquisition, supervision. N.A.K.: experimentation, visualization. H.A.: visualization. M.I.: project administration, funding acquisition. M.K.A.: visualization, high-speed imaging and processing. All authors have read and agreed to the published version of the manuscript.

Funding: This research received no external funding.

Acknowledgments: The authors acknowledge financial, technical, and administrative support provided by Pakistan Institute of Engineering and Applied Sciences for the present study.

Conflicts of Interest: The authors declare no conflict of interest.

References

1. Chun, M.H.; Kim, Y.S.; Park, J.W. An investigation of direct condensation of steam jet in subcooled water. *Int. Commun. Heat Mass Transf.* **1996**, *23*, 947–958. [CrossRef]
2. Petrovic, A. Analytical study of flow regimes for direct contact condensation based on parametrical investigation. *J. Press. Vessel. Technol.* **2005**, *127*, 20–25. [CrossRef]
3. Kazuyuki, T.; Yasuo, O.; Tomoaki, K. Numerical study on direct-contact condensation of vapor in cold water. *Fusion Eng. Des.* **2002**, *63–64*, 421–428. [CrossRef]
4. Kim, Y.S.; Park, J.W.; Song, C.H. Investigation of the steam-water direct contact condensation heat transfer coefficients using interfacial transport model. *Int. Commun. Heat Mass Transf.* **2004**, *31*, 397–408. [CrossRef]
5. Kim, Y.K.; Song, C.H. Overall review of steam jet condensation in a next generation reactor water pool. In Proceedings of the 2003 ASME International Mechanical Engineering Congress, Washington, DC, USA, 15–21 November 2003.
6. Xu, D.; Ye, Y.; Chen, Q.; Chen, G. Condensation regime diagram for supersonic and subsonic steam jet condensation in water flow in a vertical pipe. *Appl. Therm. Eng.* **2018**, *130*, 62–73. [CrossRef]
7. Datta, P. Experimental investigation on the effect of initial pressure conditions during steam-water direct contact condensation in a horizontal pipe geometry. *Int. Commun. Heat Mass Transf.* **2021**, *121*, 105082. [CrossRef]
8. Barna, I.F.; Ezsöl, G. Multiple condensation induced water hammer events, experiments and theoretical investigations. *Kerntechnik* **2011**, *76*, 231–236. [CrossRef]
9. Barna, I.F.; Imre, A.R.; Baranyai, G.; Ézsöl, G. Experimental and theoretical study of steam condensation induced water hammer phenomena. *Nucl. Eng. Des.* **2010**, *240*, 146–150. [CrossRef]
10. Xu, Q.; Liang, L.; She, Y.; Xie, X.; Guo, L. Numerical investigation on thermal hydraulic characteristics of steam jet condensation in subcooled water flow in pipes. *Int. J. Heat Mass Transf.* **2022**, *184*, 122277. [CrossRef]

engineering proceedings

Proceeding Paper

Computational Investigation of the Flow Structure through an Over-Expanded Nozzle †

Muhammad Haroon [1,*], Shummaila Rasheed [1], Muhammad Irfan [1], Manzar Masud [1] and Muhammad Arsalan Munawar [2]

1 Department of Mechanical Engineering, Capital University of Science and Technology (CUST), Islamabad 44000, Pakistan
2 Department of Mechanical Engineering, International Islamic University, Islamabad 44000, Pakistan
* Correspondence: muhammad.haroon@cust.edu.pk
† Presented at the 2nd International Conference on Advances in Mechanical Engineering (ICAME-22), Islamabad, Pakistan, 25 August 2022.

Abstract: Flow separation is a complex phenomenon that occurs in many internal and external flows. In internal flows, flow separation produces so-called side loads that are undesirable. This study aims to investigate the effect of the nozzle-pressure ratio on flow structures in a non-axisymmetric sub-scale two-dimensional (2D) convergent divergent type and in a three-dimensional axisymmetric nozzle, computationally. State-of-the-art ANSYS CFX software is used for the numerical flow analysis at two different pressure ratios: 3.0 and 3.4. Computational analysis shows that the flow is dominated by induced shock-wave boundary-layer separation. The computational results are in good agreement with the available experimental data. A considerable difference between the flow structure is observed from 3.0 to 3.4 NPRs.

Keywords: flow separation; over expanded; side loads; supersonic nozzle

Citation: Haroon, M.; Rasheed, S.; Irfan, M.; Masud, M.; Munawar, M.A. Computational Investigation of the Flow Structure through an Over-Expanded Nozzle. *Eng. Proc.* **2022**, *23*, 7. https://doi.org/10.3390/engproc2022023007

Academic Editors: Mahabat Khan and M. Javed Hyder

Published: 20 September 2022

1. Introduction

Nozzle-flow separation is a natural gas dynamics phenomenon that occurs in supersonic convergent divergent nozzles when they operate below its design nozzle pressure ratio (NPR), also called operation in over-expanded conditions. Over expanded operating conditions may occur during startup and throttling processes. It seems in the first instance that there is an overexpansion of the nozzle's enhanced thrust efficiency and control mass flow rate, but the resultant flow separation creates some serious issues for the designing of over-expanded nozzles [1]. Fluid adjusts itself inside the nozzle to fulfil the exit boundary conditions (pressure) and forms Mack disks; incident shocks; shock-wave interactions; shock-wave boundary layer interactions (SWBLIs), and also changes the entire internal geometry of the nozzle [2]. SWBLIs are the most undesirable phenomenon in nozzle designs. SWBLIs create side loads that not only damage the nozzle, but also place life-limiting constraints on the nozzle design [3].

Flow separation and its related phenomena are of great interest to researchers. Based on these phenomena, various analytical, computational, and experimental studies on sub-scale and full-scale models have been conducted [4]. Early studies reported that flow separation from the wall occurs due to the fluid's viscous effect and high pressure gradient across the flow when the nozzle operates at a low nozzle-pressure ratio (NPR) and with a comparatively large expansion ratio [1]. Later on, it was determined that there are two types of flow regimes existing in flow separation: one is called FSS (free shock separation), in which flow is completely detached from the nozzle and never reattached to the wall boundary. The second is called RSS (restricted shock separation) shown by a recirculatory bubbles on the nozzle wall where flow is first detached downstream of the nozzle and then reattached to the wall. In 1970, during the SSME (space shuttle main engine) experiment, the largest side-load was produced in the transition of FSS to RSS, while both flow regimes occurred at different nozzle-pressure ratios [3].

In the present study, the investigation of the flow structure using the geometric profile of an experimental sub-scale test CD-type nozzle is performed by considering two computation scenarios, namely, a 2D and 3D axisymmetric simulations. The main objective is to computationally analyze the flow structure of an over-expanded nozzle at higher pressure ratios.

2. Methodology

2.1. Geometry and Meshing

The experimental test nozzle employed in the current analysis was a non-axisymmetric; subscale; two-dimensional CD-type nozzle having a nominal throat area (At = 4.317 in2); expansion ratio (Ae/At = 1.797); and a constant width of 3.99 in. By keeping in mind the one-dimensional theory, the nozzle was designed with NPR = 8.78 and exit Mach number (Me) = 2.07, as presented in Figure 1a. By using the same profile of the geometry, computation was performed on the axisymmetric three- and two-dimensional nozzles. The geometry, mesh with quadrilateral elements, and domain of a 1D axisymmetric convergent–divergent nozzle are presented in Figure 1b. Moreover; Figure 2a shows the 3D axisymmetric geometry and Figure 2b shows the 3D axisymmetric mesh.

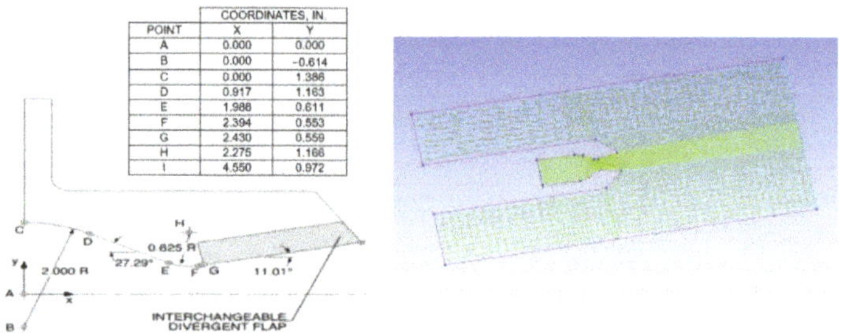

(a) (b)

Figure 1. (a) Geometry of 1D axisymmetric nozzle and (b) 2D geometry, mesh, and domain of an axisymmetric convergent–divergent nozzle.

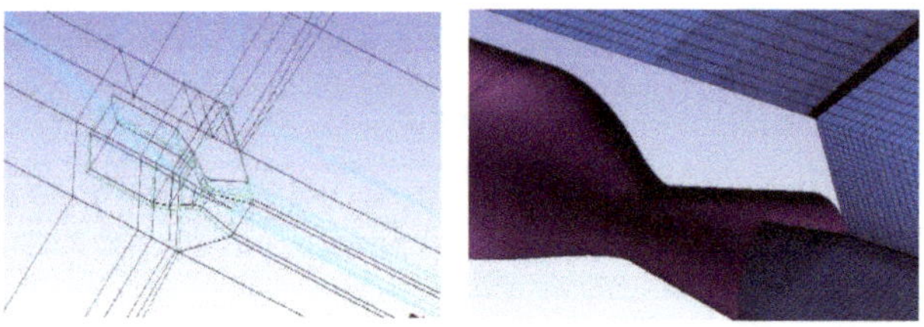

(a) (b)

Figure 2. (a) Three-dimensional axisymmetric geometry and (b) 3D axisymmetric mesh.

2.2. Computational Methods

Primarily, governing equations (Equations (1)–(4)) were transformed from the physical to computational domain by generalizing the method of co-ordinate transformation to increase the correctness of the numerical technique and implementation of boundary conditions. Later, the control volume technique was used to discretize these equations. The 2nd-order upwind scheme was employed to discretized the convective terms in the equations, whereas 2nd-order flux splitting was employed to drive inviscid fluxes to achieve the required up-winding and dissipation in the vicinity of the shock waves. The interface flux, which depends upon the upstream and downstream of the face, was determined by the separate terms. Diffusion terms were discretized using central differencing. The least-square cells technique was implemented to rebuild the variable gradients. A TVD slope limiter with the integration of the Minmod function was used, which limited the overshoots/undershoots on the cell faces. To control the numerical stiffness running at a low Mach number, a block Gauss-Seidel algorithm was solved in a coupled way to discretize the whole system. For the time derivatives, an advanced form of the 2nd-order Euler backward scheme, which is an implicit multistage time-stepping scheme, was used to obtain the physical time. On the other hand, for inner iteration, an implicit pseudo-time marching scheme was used.

$$\partial \rho/\partial t + \partial(\rho u_i)/\partial x_i = 0; \text{ (Equation of Continuity)} \quad (1)$$

$$\partial(\rho u_i)/\partial t + \partial(\rho u_i u_j)/\partial x_j = -\partial p/x_i + \partial \tau_{ij}/\partial x_j; \text{ (Momentum Equation)} \quad (2)$$

$$\partial(\rho E)/\partial t + \partial(u_i(\rho E + p)/\partial x_i = \vec{\nabla}.(\alpha_{eff}\partial T/\partial x_i + u_j(\tau_{ij}); \text{ (Energy Equation)} \quad (3)$$

$$\rho = p/RT; \text{ (Gas Law)} \quad (4)$$

In Equations (1)–(4), E represents the total energy $((h-p)/\rho + (u^2+v^2)/2))$; ρ is the density (kg/m^3); u is the x-component of the velocity; t is the time (s); x is the location in the x direction; τ is the viscous stress tensor; p is the static pressure (kPa); R is the universal gas constant (J/kg K); and T is the temperature in kelvin.

2.3. Domain and Boundary Conditions

In this research, numerical analysis was performed on 3D and 2D symmetric conditions. Nozzle topology with the grid is presented in Figures 1 and 2. Downstream and upper boundaries were located at the heights of 40 and 100 in from the nozzle-throat section. The size of the domain was free of flow reflection from the boundaries. The NPR (ratio of total inlet pressure to the back pressure) was oscillated sinusoidally, whereas the back pressure was maintained at 1.3 kPa.

2.4. Initial and Boundary Conditions

At the boundary entrance of the nozzle, the stagnation temperature and pressure were considered as the physical boundary conditions and the remaining parameters were extracted by applying numerical boundary conditions using the Riemann invariant. Furthermore, the x and k values were specified as 61,979.31/s and 1.22 m^2/s^2 at the inlet domain, respectively. The outlet boundary conditions were restricted with exit-pressure boundary conditions. The adiabatic wall and non-slip boundary conditions were employed at the solid boundaries. The wall pressure achieved from a zero-pressure gradient acted normally on the nozzle-body surfaces. The inlet temperature of the nozzle was adjusted at 298.15 K.

3. Results and Discussions

Experimental normalized centerline static pressures (p/p_0) vs. nondimensionalized stream-wise locus relative to the throat of the nozzle are presented in Figure 3a,b. The results are the representations of a classical, experimental CD-type nozzle flow and three-dimensional axisymmetric CD-type nozzle. For NPRs of 3.4 and 3.0, pressure data points indicate an internally choked, over-expanded flow producing a frail shock near the nozzle throat. The shock downstream flow started to recover at an ambient pressure (p/p_0 = 1/NPR) in a continuous pulsating fashion in an experimental nozzle. The focused schlieren flow visualization of the experimental results obtained at NPR = 3.4 are presented in Figure 4a,b. This represents a frail, oblique, downstream shock at the nozzle throat having a lambda footprint structure. However, this study at NPRs of 3.0 and 3.4 presented an oblique shock near the throat and expanding downstream. The pressure adjusted itself to an ambient pressure in a pulsating manner. The difference between the present and experimental works is clear from comparing the shock locations and boundary-layer separation points. Due to a large surface area compared to the actual experimental nozzle, the near-wall viscous effects of the flow were the dominant cause of the shock-wave boundary-layer separation and its interactions with shocks.

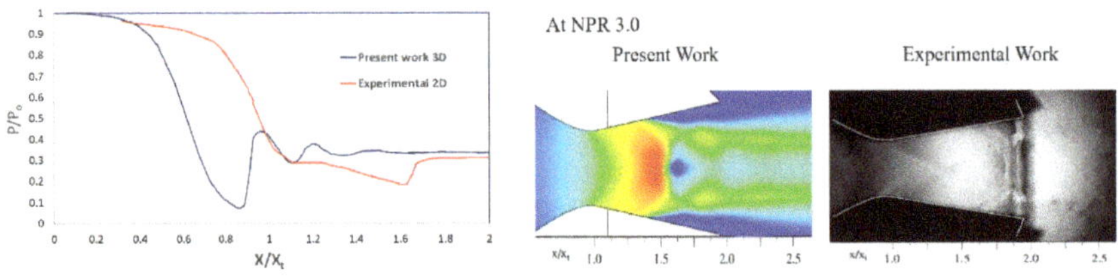

(**a**) (**b**)

Figure 3. (**a**) Graphical and (**b**) visual comparisons between the present study and the schlieren images of the experimental study at NPR = 3.

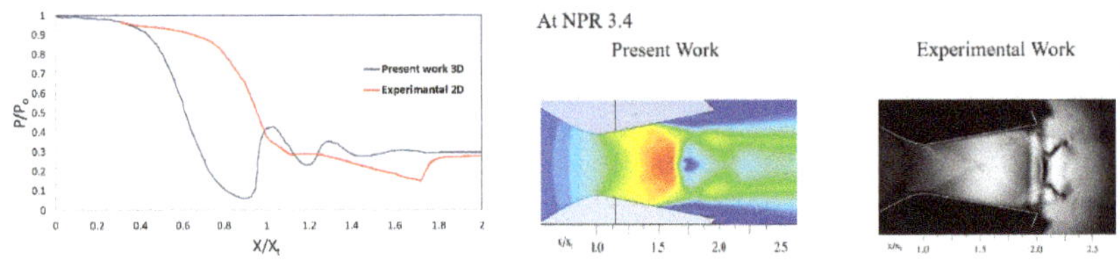

(**a**) (**b**)

Figure 4. (**a**) Graphical and (**b**) visual comparisons between the present study and the schlieren images of the experimental study at NPR = 3.4.

4. Conclusions

A computational study on an axisymmetric convergent–divergent nozzle was conducted in this research. A shock-induced boundary-layer-separation phenomenon occurred due to the dominant viscous effects observed in an axisymmetric three-dimensional nozzle in the design of over-expanded conditions during supersonic flow. A fully detached flow can clearly be seen in Figures 3 and 4 at NPRs of 3.0 and 3.4, respectively. In the axisymmetric three-dimensional CD-type nozzle, a flow-separation point occurred in half of the divergent portion as compared to the non-axisymmetric two-dimensional CD-type nozzle, because of its tendency for flow separation at supersonic speed due to viscous effects.

Author Contributions: Conceptualization; M.H. and S.R.; methodology; S.R.; software; S.R.; validation; S.R., M.I. and M.M.; formal analysis; S.R.; investigation; M.I.; data curation; M.M. and M.A.M.; writing—original draft preparation; S.R.; writing—review and editing; M.H. and M.A.M.; supervision; M.I. All authors have read and agreed to the published version of the manuscript.

Funding: This research received no external funding.

Institutional Review Board Statement: Not applicable.

Informed Consent Statement: Not applicable.

Data Availability Statement: Not applicable.

Conflicts of Interest: The authors declare no conflict of interest.

References

1. Lawrence, R.A. *Symmetrical and Unsymmetrical Flow Separation in Supersonic Nozzles*; NASA: Washington, DC, USA, 1967.
2. Hunter, C.A. Experimental Investigation of Separated Nozzle Flows. *J. Propuls. Power* **2004**, *20*, 527–532. [CrossRef]
3. Nave, L.; Coffey, G. Sea Level Side Loads in High-Area-Ratio Rocket Engines. In Proceedings of the AIAA/SAE 9th Propulsion Conference, Las Vegas, NV, USA, 5 November 1973.
4. Chen, C.L.; Chakravarthy, S.R.; Hung, C.M. Numerical Investigation of Separated Nozzle Flows. *AIAA J.* **1994**, *32*, 1836–1843. [CrossRef]

engineering proceedings

MDPI

Proceeding Paper

Mathematical Modelling to Predict the Best Inclination Angle for Maximum Distillate Output of A Solar Still †

Nawaf Mehmood Malik *, Mansoor Ali Zaheer, Asif Ali and Abdul Haseeb

Department of Mechanical Engineering, College of Engineering and Technology, University of Sargodha, Sargodha 40100, Punjab, Pakistan

* Correspondence: nawaf998428@gmail.com

† Presented at the 2nd International Conference on Advances in Mechanical Engineering (ICAME-22), Islamabad, Pakistan, 25 August 2022.

Abstract: Solar stills are generally used to obtain fresh and clean water from saline water sources using solar energy. This technique is very economical, as only the source of energy is solar energy. Many factors affect the efficiency of the solar stills, such as altitude, location, wind velocity, thickness, the inclination angle of the glass, etc. In this paper, mathematical modelling of three different solar stills with glass inclination angles of 15°, 30°, and 45° has been carried out to calculate the total radiation falling on these solar stills in addition to the calculation of the total heat transfer inside of them. These parameters are further utilized to compute the distilled water output from 10 a.m. to 4 p.m. by considering the location of the Sargodha, Punjab, Pakistan, on the 22nd of June.

Keywords: heat transfer; inclination angle; solar still

check for updates

Citation: Malik, N.M.; Zaheer, M.A.; Ali, A.; Haseeb, A. Mathematical Modelling to Predict the Best Inclination Angle for Maximum Distillate Output of A Solar Still. *Eng. Proc.* **2022**, 23, 5. https://doi.org/10.3390/engproc2022023005

Academic Editors: Mahabat Khan, M. Javed Hyder, Muhammad Irfan and Manzar Masud

Published: 20 September 2022

1. Introduction

A lot of the solar desalination plants that are operational today have a major drawback in that they are very costly and utilize a lot of energy [1]. Scientists have predicted that the demand for water will be 56% more than its supply by 2025 [2]. One of the most reliable technologies to solve water scarcity issues is water desalination using solar stills [3]. Factors that affect the yield of solar stills are solar intensity, the temperature of the air and its velocity, differences in temperature between the glass and water vapor, the water surface area, water depth, the inlet water temperature, the area of the basin, and the inclination angle of the glass [4].

Researchers have shown that for solar distillation, the most suitable material for these covers is glass, as it has higher solar radiation transmittance at different angles [5]. A 4 mm glass cover thickness is considered the best [6]. A 2 cm water depth is the best for producing the most distilled water using solar stills [7].

A higher water temperature inside the solar still will result in increased distilled water output [8]. The productivity of the solar still also depends upon the surface area of the collector [9]. The inclination angle of the glass surface and the direction of the glass cover on the still depend on the latitude of the specified location [10].

2. Methodology

The three different angles of inclination being used are 15°, 30°, and 45°. The total radiation falling on the solar stills is calculated theoretically; then, heat transfer calculations of the solar still are carried out to find the particular inclination angle on which maximum yield occurs. The calculations were conducted for a time period of 6 h on 22 June 2021 for the location of Sargodha city. The following assumptions were made while conducting the calculations [11]:

In the solar still, there is no leakage of vapors, and there is no heat loss in the basin.

1. There is a constant water level maintained in the basin.
2. Only filmwise condensation occurs, and it occurs in place of dropwise condensation.

Filmwise condensation is considered because in dropwise condensation, the transfer rates of heat are more than ten times those of film-type condensation [12]. The design of the solar stills with the inclination angles of 15°, 30° and 45° are given in Figure 1:

Figure 1. Solar stills with different inclination angles.

The area of the water surface is 0.403 m^2. The depth of the water is kept at 2 cm or 20 mm for best performance [1]. The thickness of the glass is kept at 4 mm for the best results [8], and the thickness of the base metal is 6mm. Glass is used for the cover material [7]. The calculations for total radiation on the solar still and heat transfer are for the passive single-slope solar still model.

3. Total Radiations Calculations on Solar Still

We can find the radiation (I_t) available to the solar still as follows:

$$I_t = [\, I_b\, R_b + I_d \left(\frac{1+\cos\beta}{2}\right) + I_{Pg}(\frac{1-\cos\beta}{2})] \tag{1}$$

The amount of solar radiation available above the surface of the Earth (I_0) is:

$$I_0 = \frac{12 * 3600}{\pi} Gsc \left(1 + 0033\cos\frac{360n}{365}\right) \left[\cos\phi\cos\,\delta\,(\sin\,\omega2 + \sin\,\omega1) + \frac{\pi(\omega2 -\ \omega1)}{180}\,(\sin\,\phi\,\sin\,\delta)\right] \tag{2}$$

The incident angle of sun rays for tilted surfaces is θ. It can be computed as:

$$\begin{aligned} \theta = \ & (\sin(\phi)\ \cos(\beta)\ \sin(\delta) - \cos(\phi)\ \cos(\gamma)\ \sin(\beta)\ \sin(\delta) + \cos(\phi)\ \cos(\omega)\ \cos(\beta) \\ & \cos(\delta) + \sin(\phi)\ \cos(\omega)\ \sin\ \cos(\gamma)\ \cos(\delta) + \sin(\beta)\ \sin(\omega)\ \sin(\gamma)\ \cos(\delta))\ \cos^{-1} \end{aligned} \tag{3}$$

The diffuse fraction is calculated using the model of Orgill and Holland:

$$\text{For } 0 < K_t < 0.35 \ \text{ when } I_d/I = (1 - (K_t)\ 0.249) \tag{4}$$

$$\text{For } 0.35 \le K_t \le 0.75 \ \text{ when } I_d/I = (1.577 - (K_t)\ 1.84) \tag{5}$$

$$\text{For } 0.75 < K_t \le 1 \ \text{ when } I_d/I = (0.177) \tag{6}$$

The beam radiation tilt factor (R_b) is given as:

$$R_b = \frac{\text{Cos}\theta}{\text{Cos}\theta_z} \tag{7}$$

4. Heat Transfer Calculations of the Solar Still

The convective transfer of heat between glass and water (Q_{cwg}) is calculated as follows:

$$Q_{cwg} = 0.884[(T_w - T_g) + \frac{P_w - P_g}{268e^3 - P_w} \times T_w\ (T_w - T_g)] \tag{8}$$

The water temperature in the water basin (T_w) and glass temperature is calculated as:

$$T_w = \frac{f(t)}{\alpha}(1 - \exp(-\alpha_w \tau_w))\exp(-\alpha_w \tau_w) \quad (9)$$

Here, the emittance of glass is kept at 0.05 [13]. The water heat capacity present in the basin is 4.184 J/K [14]. The absorptivity of the glass is 0.05 [15]. The absorptance of the base plate is 0.92 [16]. The temperature of the glass (T_g) can be calculated using the following formulas given below:

$$T_g = \frac{(0.022612{T_w}^2 - 15.76T_w + 2392)*T_w*h_w*T_a*Ar + Ar(0.048T_a - 9)T_s}{(0.022612{T_w}^2 - 15.76*T_w + 2392)*h_w*T_a*Ar + (0.040T_a - 9)} \quad (10)$$

To calculate the temperature of the base plate (T_b) or water basin, use the given Equation:

$$T_b = \frac{\alpha_b \tau_g \tau_w\, I(t) + h_{cblw} \times T_w + U_b \times T_a}{h_{cblw} + U_b} \quad (11)$$

The evaporative mass transfer to the glass cover (m_{ev}) from the water surface is calculated as:

$$M_d = 3600 \int_{t_1}^{t_2} M_{ev}\, dt \quad (12)$$

The solar water desalination unit's thermal efficiency (η) can be calculated as

$$\eta = \frac{q_{ewg}}{I_t} \quad (13)$$

5. Results and Discussion

Solar Still with glass inclination angle of 45° gives maximum distilled output as compared to its counter parts in Sargodha region on 22 June. Its productivity is 12.41% more than the solar stills with an inclination angle 30° and 13.35% more than the solar still with an inclination angle of 15°, as shown in Tables 1–3.

Table 1. Time, total radiation, and distillate water for 15° solar still.

Time (Hours)	Total Radiations It (W/m^2)	Distillate Water (Liters)
10–11 a.m.	1200.121	0.539977
10–12 p.m.	1199.938	1.197339
12–1 p.m.	1199.994	1.992371
1–2 p.m.	1200.184	2.950399
2–3 p.m.	1200.212	3.320589
3–4 p.m.	1200.043	3.592003

Table 2. Time, total radiation, and distillate water for 30° solar still.

Time (hours)	Total Radiations It (W/m^2)	Distillate Water (Liters)
10–11 a.m.	1256.421	0.585175
10–12 p.m.	1256.443	1.305578
12–1 p.m.	1256.436	2.190179
1–2 p.m.	1256.411	3.277347
2–3 p.m.	1256.407	3.630326
3–4 p.m.	1256.431	3.916858

Table 3. Time, total radiation, and distillate water for 45° solar still.

Time (hours)	Total Radiations It (W/m^2)	Distillate Water (Liters)
10–11 a.m.	1176.705	0.993127
10–12 p.m.	1176.471	1.477627
12–1 p.m.	1176.551	2.303178
1–2 p.m.	1176.776	3.465561
2–3 p.m.	1176.806	3.838896
3–4 p.m.	1176.613	4.606433

6. Conclusions

In this paper, we applied numerical modelling techniques to calculate the distillate water output of passive solar stills with three different inclination angles, i.e., 15°, 30°, and 45°. The location of Sargodha, Punjab, Pakistan, on 22 June was the considered context. Results were calculated after every hour of operation from 10 a.m. to 4 p.m. After applying the mathematical modelling technique, it was shown that solar stills with a 45° angle achieve the maximum amount of distillate, which was 4.606433 L more than the solar stills with 15° and 30° inclination angles, which had distillate outputs of 3.592003 L and 3.916858 L, respectively. A proven mathematical technique used to find the distillate water output of passive solar stills was proposed by Bao Nguyen [17], and we have further utilized this method to predict the best inclination angle for maximum passive single-slope solar still productivity.

Author Contributions: Conceptualization, N.M.M. and M.A.Z.; methodology, N.M.M. and A.H.; software, N.M.M. and A.A.; validation, N.M.M. and A.A.; formal analysis, N.M.M. and M.A.Z.; investigation, N.M.M. and A.H.; data curation, M.A.Z.; writing—original draft preparation, N.M.M.; writing—review and editing, N.M.M. and A.H.; supervision, A.H.; project administration, N.M.M. All authors have read and agreed to the published version of the manuscript.

Funding: This research received no external funding.

Institutional Review Board Statement: Not applicable.

Informed Consent Statement: Not applicable.

Data Availability Statement: Not applicable.

Conflicts of Interest: The authors declare no conflict of interest.

References

1. Chandrashekara, M.; Yadav, A. Water Desalination System Using Solar Heat: A Review. *Renew. Sustain. Energy Rev.* **2017**, *67*, 1308–1330. [CrossRef]
2. Shatat, M.; Worall, M.; Riffat, S. Opportunities for Solar Water Desalination Worldwide: Review. *Sustain. Cities Soc.* **2013**, *9*, 67–80. [CrossRef]
3. Kabeel, A.E.; Abdelgaied, M.; Mahmoud, G.M. Performance Evaluation of Continuous Solar Still Water Desalination System. *J. Therm. Anal. Calorim.* **2021**, *144*, 907–916. [CrossRef]
4. Sharshir, S.W.; Ellakany, Y.M.; Algazzar, A.M.; Elsheikh, A.H.; Elkadeem, M.R.; Edreis, E.M.A.; Waly, A.S.; Sathyamurthy, R.; Panchal, H.; Elashry, M.S. A Mini Review of Techniques Used to Improve the Tubular Solar Still Performance for Solar Water Desalination. *Process Saf. Environ. Prot.* **2019**, *124*, 204–212. [CrossRef]
5. Sarkar, M.N.I.; Sifat, A.I.; Reza, S.M.S.; Sadique, M.S. A Review of Optimum Parameter Values of a Passive Solar Still and a Design for Southern Bangladesh. *Renew. Wind. Water Sol.* **2017**, *4*, 1. [CrossRef]
6. Panchal, H. Performance Investigation on Variations of Glass Cover Thickness on Solar Still: Experimental and Theoretical Analysis. *Technol. Econ. Smart Grids Sustain. Energy* **2016**, *1*, 7. [CrossRef]
7. Keshtkar, M.; Eslami, M.; Jafarpur, K. Effect of Design Parameters on Performance of Passive Basin Solar Stills Considering Instantaneous Ambient Conditions: A Transient CFD Modeling. *Solar Energy* **2020**, *201*, 884–907. [CrossRef]
8. Panchal, H.; Mevada, D.; Sathyamurthy, R. The Requirement of Various Methods to Improve Distillate Output of Solar Still: A Review. *Int. J. Ambient. Energy* **2021**, *42*, 597–603. [CrossRef]
9. Durkaieswaran, P.; Murugavel, K.K. Various Special Designs of Single Basin Passive Solar Still—A Review. *Renew. Sustain. Energy Rev.* **2015**, *49*, 1048–1060. [CrossRef]

10. Goshayeshi, H.; Safaei, M.R. Effect of Absorber Plate Surface Shape and Glass Cover Inclination Angle on the Performance of a Passive Solar Still. *Int. J. Numer. Methods Heat. Fluid Flow.* **2019**. ahead-of-print. [CrossRef]
11. Panchal, H.; Shah, P. Modelling and Verification of Single Slope Solar Still Using ANSYS-CFX. *Int. J. Energy Environ.* **2011**, *2*, 985–988.
12. Çengel Yunus, A.; Ghajar, A.J. *Heat and Mass Transfer: Fundamentals & Applications*, 4th ed.; McGraw-Hill: New York, NY, USA, 2011.
13. Dev, R.; Tiwari, G.N. Characteristic Equation of a Passive Solar Still. *Desalination* **2009**, *245*, 246–265. [CrossRef]
14. Ferrer, G.; Barreneche, C.; Solé, A.; Martorell, I.; Cabeza, L.F. New Proposed Methodology for Specific Heat Capacity Determination of Materials for Thermal Energy Storage (TES) by DSC. *J. Energy Storage* **2017**, *11*, 1–6. [CrossRef]
15. Monteiro, J.; Baptista, A.; Pinto, G.; Ribeiro, L.; Mariano, H. Assessment of the Use of Solar Desalination Distillers to Produce Fresh Water in Arid Areas. *Sustainability* **2019**, *12*, 53. [CrossRef]
16. Bergström, D.; Powell, J.; Kaplan, A.F.H. The Absorptance of Steels to Nd:YLF and Nd:YAG Laser Light at Room Temperature. *Appl. Surf. Sci.* **2007**, *253*, 5017–5028. [CrossRef]
17. Bao, N.T. The Mathematical Model of Basin-Type Solar Distillation Systems. In *Distillation*; Steffen, V., Ed.; IntechOpen: Rijeka, Croatia, 2019.

engineering proceedings

MDPI

Proceeding Paper

Design and Simulation of Thermoelectric Generator to Enhance the Cooling Rate and Power Generation from Waste Heat of Chimney by Employing Different Angles of Flaps †

Asif Ali, Mansoor Ali Zaheer and Nawaf Mehmood Malik *

Department of Mechanical Engineering, College of Engineering and Technology, University of Sargodha, Sargodha 40100, Punjab, Pakistan
* Correspondence: nawaf998428@gmail.com

† Presented at the 2nd International Conference on Advances in Mechanical Engineering (ICAME-22), Islamabad, Pakistan, 25 August 2022.

Abstract: This study consists of a modern and economic way to recover the heat wasted from the industrial chimneys which are cooled down by natural sources by using thermal electric generator heat sinks. A flap with altered heat sink (which is attached at the top of thermo-electric generator) is used in place of convectional heat sink base. A 3D model is proposed by using AUTODESK Fusion 360 and is solved by using Workbench 2021 R1 ANSYS. The presented setup fully describes the transfer of heat along the one vertical bar inside the TEGs module which is mounted along the vertical wall of the Chimney. The impact of Flap dimensions (Height, Depth, and Angle) and conductive material performance is studied. The flap angles are 45°, 50° and 60° and depths of the flap are 28 mm, 30 mm and 33 mm. This altered heat sink accomplishes about 28% enhance in the rate of cooling of TEGs module. The maximum output power, i.e., 105 mV of the TEGs module is at 60° and at 33 mm depth. The results show that remodified heat sink maintains the system simple and requires less maintenance and also improves the cooling rate of the thermoelectric generator, which as a result improves its performance and make it reliable.

Keywords: thermoelectric generator; cooling rate; power generation; waste heat; modified

Citation: Ali, A.; Zaheer, M.A.; Malik, N.M. Design and Simulation of Thermoelectric Generator to Enhance the Cooling Rate and Power Generation from Waste Heat of Chimney by Employing Different Angles of Flaps. *Eng. Proc.* **2022**, *23*, 21. https://doi.org/10.3390/engproc2022023021

Academic Editors: Mahabat Khan, M. Javed Hyder, Muhammad Irfan and Manzar Masud

Published: 21 September 2022

1. Introduction

The operation of the TEG is dependent upon Seebeck effect, which is used to generate electrical power generation by applying significant temperature difference over the minimal thickness. Lv et al. [1] suggested different techniques for the finer cooling of the TEGs. Demir and Dincer proposed shell and tube type heat exchanger for the removal of exhaust fumes from the exhaust, while fresh air is used to cool the TEGs module. Maximum production of power from this system is 158 W correlate with the density of power 108.8 W/m^2 [2]. Figure 1 shows (a) TEG module domain and its (b) Geometry.

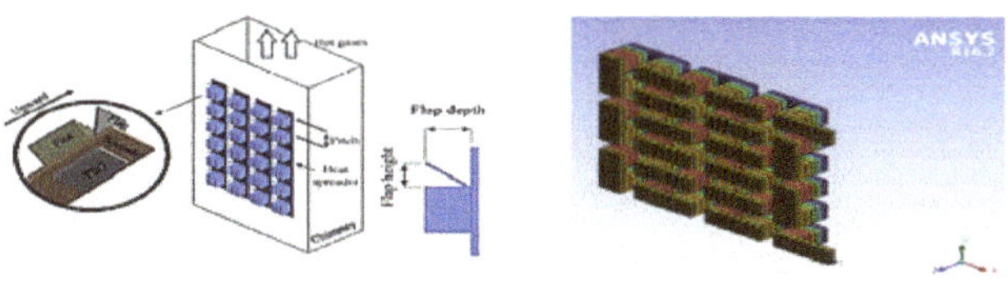

Figure 1. (**a**) TEG Module Domain (**b**) Geometry of TEG.

Only limited research has been carried out on the TEG for recovery of dissipated heat from a smokestack when it subjected to passive cooling. Ansys fluent module is used to solve a 3D CFD model described in this paper. The heat is deviated away from the chimney by employing flaps on the upper side of the heat sink. Flaps are angled at various angles and have varying depths. The 3D model represents all the features of the temperature distribution, directional electric field strength, potential difference, and total heat flow rate. The goal of the study is to recuperate the heat dissipated from the funnel and transfer this wasted heat into electrical power.

2. Methodology

2.1. Physical Domain

Flap used has a specific inclination angles 'α' and specific length 'L_f' in respect of the upright surface of the chimney. Flap contiguity with heat abstraction thickness of base is 0.5 mm and fins of the heat sink is 2 mm. Number of fins mounted on heat sink is 13, with a height of 30 mm and its thickness is 2 mm, its height from the base of the heat sink is 30.5 mm. The purpose of these modifications in the heat sink is used to increase the heat transfer and cool the TEG's module mounted on chimney's vertical surface [3]. A heat spreader has a thickness of $75 \times 90 \times 0.5$ mm is mounted between heat sink vertical surface and wall of chimney. TEG module structure consists of 36 Thermocouples whose positive leg is made up of Bismuth Telluride (BiTe+) and negative leg is (BiTe−) having leg length 'Lg' 1.4 mm. Figure 2 shows the TEG schematic diagram.

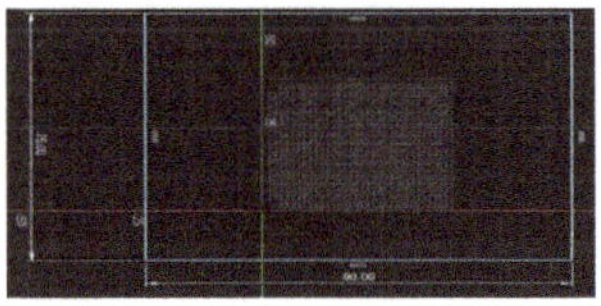

Figure 2. Schematic diagram of TEG.

2.2. Governing Equations

For solid domain, heat scattering equation is used to govern the rate of heat transfer across (TEG, spreader, and heat sink) [4]. Following equations help to govern flow of fluid and heat transfer which surround the heat absorber [5]. Equations (4)–(6), respectively, represent them [6,7].

$$\nabla.(k\vec{\nabla}T + |\vec{J}|^2/\sigma - T\frac{d\alpha}{dT}\vec{J} \cdot \vec{\nabla}T) = 0 \quad (1)$$

$$\nabla \cdot \vec{J} = 0 \quad (2)$$

$$\vec{\varphi} = \vec{J}/\sigma + \alpha\nabla T \quad (3)$$

$$\nabla\left(\rho\vec{v}\right) = 0 \quad (4)$$

$$\nabla\left(\rho\vec{v}\vec{v}\right) = -\nabla P + \rho\vec{g} \quad (5)$$

$$\nabla\left(\vec{v}(\rho E + P)\right) = \nabla(k\nabla T) \quad (6)$$

2.3. Material Properties

Figure 3 shows Material behavior of positive and negative leg. Material used is Bismuth-Telluride alloy.

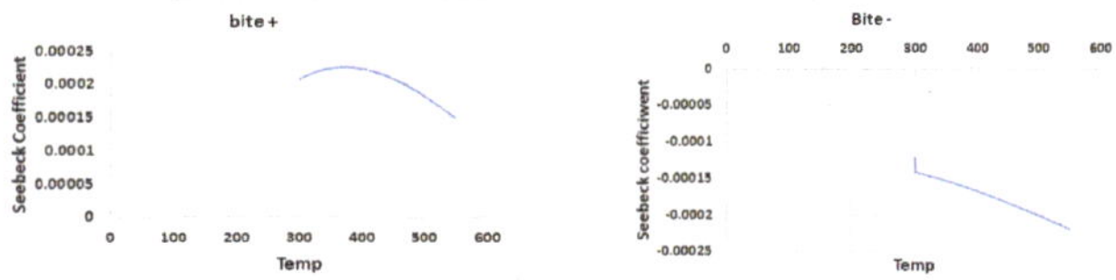

(**a**) (**b**)

Figure 3. (**a**) Material behavior of positive leg (**b**) Material behavior of negative leg.

Equations (7)–(9) show the See beck coefficient, electrical conductivity, and thermal conductivity for both legs [8,9]. Copper (thermal conductivity 385 W/mK) is used. Atmospheric Air is considered as perfect gas [10].

$$\sigma_p = 3.32 \times 10^5 - 1109.77T + 1.06T^2 \quad (7)$$

$$\sigma_n = -8822.79 + 267.66T - 0.73T^2 + 6.31 \times 10^{-4}T^3 \quad (8)$$

$$\alpha_P = -4.38 \times 10^{-4} + 4.17 \times 10^{-6}T - 8.04 \times 10^{-9}T^2 + 4.36 \times 10^{-12}T^3 \quad (9)$$

2.4. Numerical Solving

These governing equations are valid for TEG modules as long material and formation of design do not change [11]. There are considerations of the buoyancy effect in which ambient pressure varies linearly with the change of height Equation (14).

$$q_{TEG} = [W] = -2.648 \times 10^{-4}\Delta T^2 \quad (10)$$

$$[W] = -4.81 \times 10^{-13}\Delta T^5 + 2.16 \times 10^{-10}\Delta T^4 + 3.06 \times 10^{-8}\Delta T^3 + 1.92 \times 10^{-5}\Delta T^2 - 6.14 \times 10^{-5}\Delta T \quad (11)$$

$$V_{TEG} = 7.43 \times 10^{-6}\Delta T^2 + 5.29 \times 10^{-13}\Delta T \quad (12)$$

$$\eta_{TEG} = 1.79 \times 10^{-7}\Delta T^2 + 3.46 \times 10^{-5}\Delta T \quad (13)$$

$$P_{amb} = 101325[P_a] - \rho gH \quad (14)$$

3. Results and Discussion

3.1. Impact of Flap Depth

Figure 4 displays effect of temperature and voltage on depth of flap. The result shows that by using conventional heat sink with TEG module has the lowest output power and has minimum efficiency.

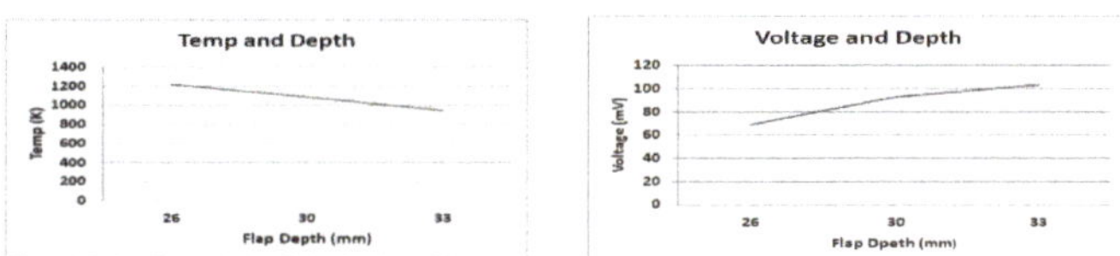

(**a**) (**b**)

Figure 4. (**a**) Impact of Temperature on depth (**b**) Impact of Voltage on depth.

3.2. By Using Conductive Flap

It is clearly shown from the Figure 5 that as we use flap to enhance the cooling rate of TEG the output voltage also increases.

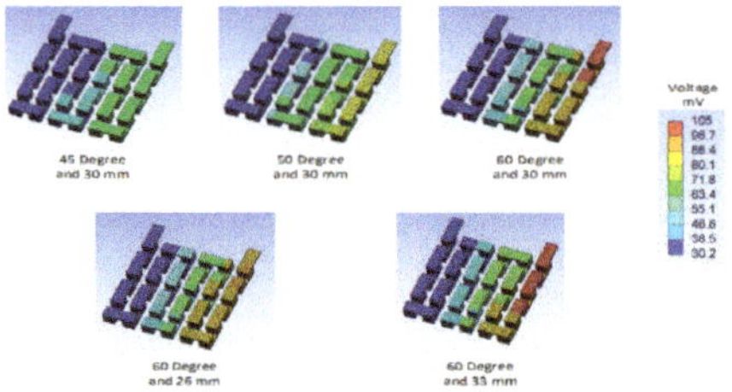

Figure 5. Variation of voltage at different angle.

At 45 degree the max voltage is 57.8 mV which is 39% more than without flap. At 50 degree the max voltage is 72.5 mV which is 20% more than at 45 degrees. At 60-degree output max voltage is 93.4 mV which is 22% more than that at 50 degrees. Now by increasing flap depth out max voltage also increases.

3.3. *Effect of Flap Angle*

From above Figure 6 it can be concluded that on smaller angle, TEG voltage is negligible. So, it is concluded that as the tilt angle of flap increases, the output power also enhances. Figure 7 shows variation of voltage on different angled flaps (i.e., 0°, 30°, 50° and 60°).

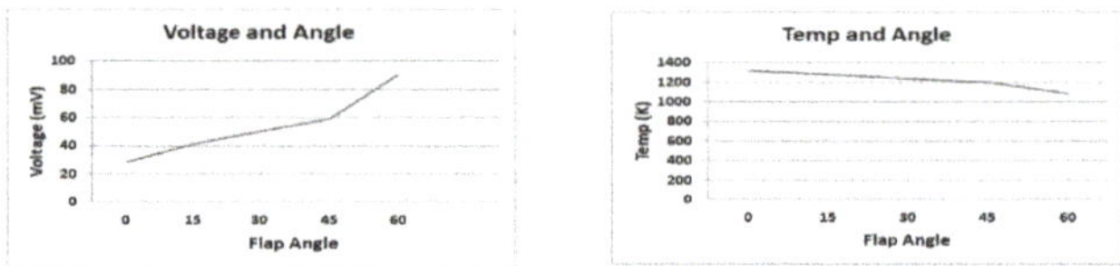

(**a**) (**b**)

Figure 6. (**a**) Effect of voltage on flap angle (**b**) Effect of temperature on flap angle.

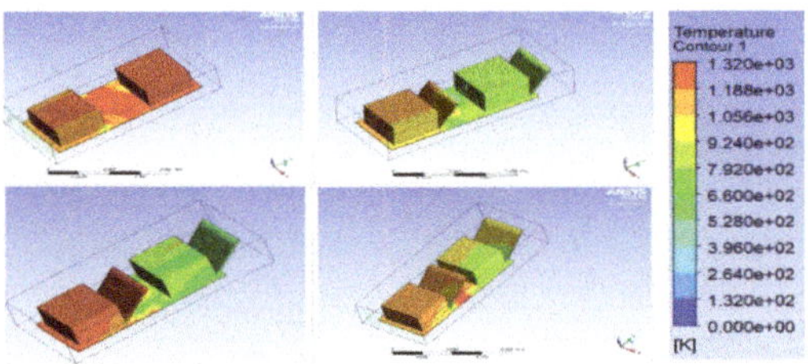

Figure 7. Variation of voltage on different angled flaps.

4. Conclusions

This study presents numerical investigation to effectively recover waste heat from the industrial sources and produce electricity using thermoelectric generator. The flap angle and the depth of the flap are varied to find the best configuration. Flap of 45°, 50° having depth of 30 mm, 30 mm, respectively, and 60° having depth of 26 mm, 30 mm and 33 mm are used. Flap with angle 600 and having depth 33 mm provides maximum cooling rate 28% and produce voltage 105 mV.

Author Contributions: Conceptualization, A.A. and M.A.Z.; methodology, A.A. and N.M.M.; software, A.A. and N.M.M.; validation, A.A. and N.M.M.; formal analysis, A.A. and M.A.Z.; investigation, A.A. and N.M.M.; data curation, M.A.Z.; writing—original draft preparation, A.A.; writing—review and editing, A.A. and N.M.M.; supervision, A.A.; project administration, A.A. All authors have read and agreed to the published version of the manuscript.

Funding: This research received no external funding.

Institutional Review Board Statement: Not applicable.

Informed Consent Statement: Not applicable.

Data Availability Statement: Not applicable.

Conflicts of Interest: The authors declare no conflict of interest.

References

1. Lv, S.; He, W.; Jiang, Q.; Hu, Z.; Liu, X.; Chen, H.; Liu, M. Study of Different Heat Exchange Technologies Influence on the Performance of Thermoelectric Generators. *Energy Convers. Manag.* **2018**, *156*, 167–177. [CrossRef]
2. Demir, M.E.; Dincer, I. Performance Assessment of a Thermoelectric Generator Applied to Exhaust Waste Heat Recovery. *Appl. Therm. Eng.* **2017**, *120*, 694–707. [CrossRef]
3. Yazicioğlu, B.; Yüncü, H. Optimum Fin Spacing of Rectangular Fins on a Vertical Base in Free Convection Heat Transfer. *Heat Mass Transfer* **2007**, *44*, 139. [CrossRef]
4. John Wiley & Sons, Ltd. Thermoelectric Generators. In *Thermoelectrics: Design and Materials*; John Wiley & Sons, Ltd.: Hoboken, NJ, USA, 2016; pp. 8–22. ISBN 9781118848944.
5. John Wiley & Sons, Ltd. Modeling of Thermoelectric Generators and Coolers with Heat Sinks. In *Thermoelectrics: Design and Materials*; John Wiley & Sons, Ltd.: Hoboken, NJ, USA, 2016; pp. 99–119. ISBN 9781118848944.
6. Hassan, H. Heat Transfer of Cu–Water Nanofluid in an Enclosure with a Heat Sink and Discrete Heat Source. *Eur. J. Mech. B Fluids* **2014**, *45*, 72–83. [CrossRef]
7. Eldesoukey, A.; Hassan, H. 3D Model of Thermoelectric Generator (TEG) Case Study: Effect of Flow Regime on the TEG Performance. *Energy Convers. Manag.* **2019**, *180*, 231–239. [CrossRef]
8. Lin, C.-C.; Ginting, D.; Lydia, R.; Lee, M.H.; Rhyee, J.-S. Thermoelectric Properties and Extremely Low Lattice Thermal Conductivity in P-Type Bismuth Tellurides by Pb-Doping and PbTe Precipitation. *J. Alloys Compd.* **2016**, *671*, 538–544. [CrossRef]
9. Yim, J.-H.; Jung, K.; Yoo, M.-J.; Park, H.-H.; Kim, J.-S.; Park, C. Preparation and Thermoelectric Properties of Quaternary Bismuth Telluride–Indium Selenide Compound. *Curr. Appl. Phys.* **2011**, *11*, S46–S49. [CrossRef]
10. Çengel Yunus, A.; Ghajar, A.J. *Heat and Mass Transfer: Fundamentals & Applications*, 4th ed.; McGraw-Hill: New York, NY, USA, 2011.
11. Klaus, H. *Jet Engines: Fundamentals of Theory, Design and Operation*; The Crowood Press: Marlborough, UK, 2010.

engineering proceedings

MDPI

Proceeding Paper

Refrigeration Potential Investigation of Liquefied Petroleum Gas under Atmospheric Conditions †

Atif Muzaffar, Muhammad Hasnain Tariq, Ahmad Abbas, Muhammad Tayyab and Taqi Ahmad Cheema *

Faculty of Mechanical Engineering, Ghulam Ishaq Khan Institute of Engineering Sciences and Technology, Topi 23460, Pakistan

* Correspondence: tacheema@giki.edu.pk

† Presented at the 2nd International Conference on Advances in Mechanical Engineering (ICAME-22), Islamabad, Pakistan, 25 August 2022.

Abstract: One of the potential refrigerants for refrigeration systems is liquefied petroleum gas (LPG) that can absorb latent heat from the surrounding and provide cooling, if introduced in liquid state. The present study determines the cooling effect produced in flowing water in coils after exchanging heat with liquid LPG, coming from an inverted cylinder. In an insulated box with a copper coil, the water flow rates were varied while maintaining the amount of surrounding liquid LPG. The results reveal that the cooling effect is proportional to the rate at which water flows, but the time for liquid LPG to evaporate decreases. For smallest water flow rates, the temperature differential across the water inlet and outlet was found to be the largest.

Keywords: liquid LPG; latent; heat of vaporization; copper coil; atmospheric conditions

check for updates

Citation: Muzaffar, A.; Tariq, M.H.; Abbas, A.; Tayyab, M.; Cheema, T.A. Refrigeration Potential Investigation of Liquefied Petroleum Gas under Atmospheric Conditions. *Eng. Proc.* **2022**, *23*, 32. https://doi.org/10.3390/engproc2022023032

Academic Editors: Mahabat Khan, M. Javed Hyder, Muhammad Irfan and Manzar Masud

Published: 23 September 2022

1. Introduction

Liquefied petroleum gas (LPG) is one of the most widely used fuels in the world, with applications ranging from heating and cooking, to automotive fuels. In addition to its stated purpose, LPG is increasingly being considered as a refrigerant, e.g., for cooling. According to recent studies, LPG has the potential to produce air-conditioning and refrigeration effects [1,2]. Because LPG has a very low saturation temperature, it absorbs latent heat from its surroundings, lowering the surrounding temperature. Both Compressed Natural Gas and LPG produce a prospective cooling effect when evaporated within the vaporizer units. When stored in a cylinder, LPG has a lower pressure than CNG, namely 0.8–1.0 MPa for LPG and 20–27 MPa for CNG. The former is easier to handle because it is under less pressure and in a liquid state [3]. LPG is a possible replacement refrigerant with a low global warming potential due to restrictions on the refrigerants that are destroying the ozone layer. Despite being highly combustible, it is widely used. The risk of flammability has been found to be reduced by the addition of modest amounts of components such as carbon dioxide [4]. LPG often contains a combination of propane, butane, or isobutene. Setiyo et al. explored, by simulation, that the composition of LPG affects the cooling impact created [5].

The current study investigates the cooling effect of LPG by exchanging heat between water flowing into a copper tube surrounded by LPG. Water flows at different flow rates in a circular copper tube that is directly in contact with LPG at atmospheric pressure and is enclosed in an insulated box. The LPG used in the experiment was composed of 65% propane and 35% butane [2]. The goal of the tests was to see if LPG has the ability to produce refrigeration in flowing water under normal atmospheric conditions. The LPG is stored as a liquid in a cylinder under high pressure. It evaporates and absorbs heat from the environment as it expands. LPG is chosen in liquid form from the cylinder and poured into an insulated box where heat is exchanged between running water in copper tubes surrounded by liquid LPG about to evaporate. This evaporation cools the surrounding and decreases the temperature. The cooling characteristics of LPG under atmospheric pressure for various water flow rates were investigated using an experimental setup designed in house.

2. Materials and Methods

Methodology and Setup Design

The experimental setup schematic is shown in Figure 1, a copper coil surrounded by liquid LPG was stored in an insulated box to examine the possibility of LPG to create cooling in flowing water. Water inlet and outlet temperatures were recorded with a waterproof LM35 sensor, and water flow rate was measured with a flow sensor connected to a computer through an Arduino Uno.

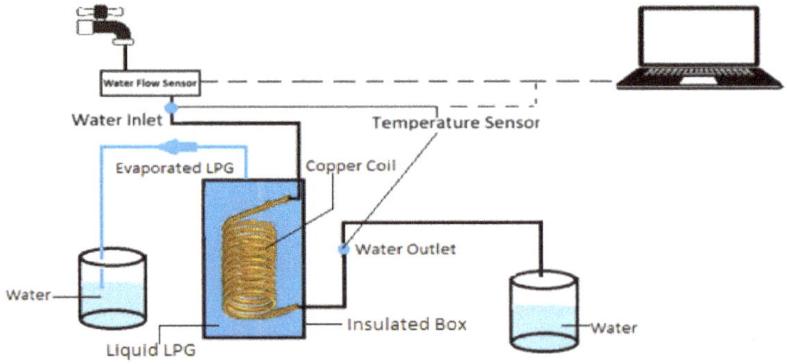

Figure 1. Schematic of Experimental Setup.

The flow of water in the copper coil absorbs the latent heat to evaporate the surrounded liquid LPG. As a result, the temperature of the water begins to drop. Equation 1 was used to calculate the rate of heat transfer.

$$\dot{Q} = \dot{m}C\Delta T \quad (1)$$

where $\dot{Q}$ is the rate heat transfer, $\dot{m}$ is the mass flow rate of water, C is the specific heat constant, and ΔT is the change in temperature of the water.

3. Design of Experiments

A regulating tap was used to control the water flow rate. The trials were carried out at four different water flow rates: 0.03 kg/s, 0.07 kg/s, 0.12 kg/s, and 0.15 kg/s. The volume of LPG in the insulated container remained constant, at 125 mL. At atmospheric circumstances, the copper coil was submerged in a container filled with LPG. Water was passed through the coil tube, which had an internal diameter of 6 mm. Temperature sensors were installed at both the intake and the outflow of the copper coil, while a water flow sensor was installed only at the inlet. For each flow rate, a predetermined volume of LPG was used. LPG evaporated as heat exchange occurred between running water and the surrounding liquid LPG. By providing passage over an insulated box, evaporated LPG vapors were dropped in the water. As the amount of Liquid LPG in the water diminished, the temperature differential between the input and output reduced, and eventually the difference disappeared as the liquid LPG evaporated.

4. Results and Discussion

Figure 2a shows that, for a low water flow rate of 0.03 kg/s, the temperature drops by 10 °C, and it takes approximately 180 s for 125 mL Liquid LPG to evaporate and the flow water to return to its initial temperature. It can also be concluded from Figure 2a–d that, as the flow rate increases, the time to evaporate the same amount of liquid LPG decreases, and the maximum decrease in water temperature was observed at low water flow rate.

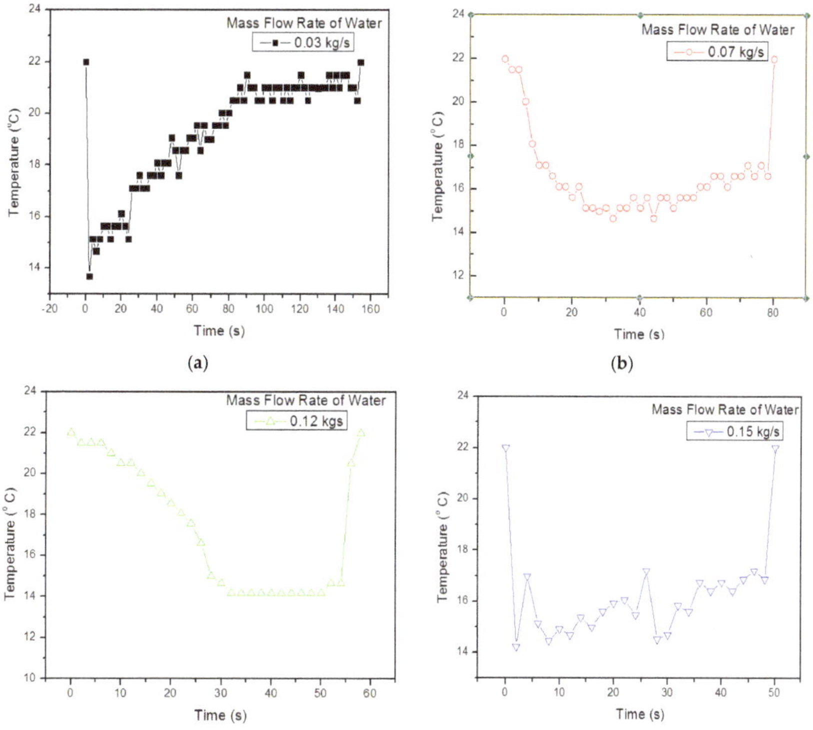

(**a**) (**b**) (**c**) (**d**)

Figure 2. Water outlet temperature as a function of time for different water flow rates.

Figure 2a–d indicate that, at low flow rates, the temperature of flowing water drops abruptly and gradually rises as the amount of liquid LPG evaporates, whereas, at higher flow rates, the temperature gradually decreases and remains nearly constant for a few seconds.

Figure 3 shows that, as the mass flow rate increases, the heat transfer rate with liquid LPG also increases. As a result of the increased heat transfer rate, the liquid-surrounded LPG evaporates faster. For example, at a flow rate of 0.15 kg/s, 125 mL liquid LPG evaporates in approximately 50 s, while at a flow rate of 0.03 kg/s, liquid LPG evaporates in approximately 160 s.

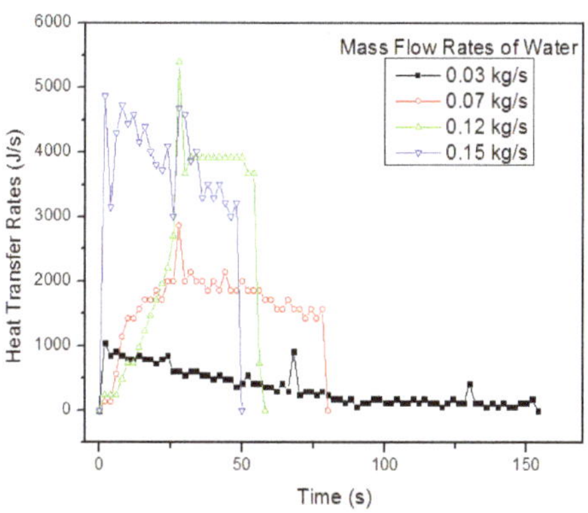

Figure 3. Heat transfer rate of water with surrounded liquid LPG at diferent water flow rates.

Flow rates influence the temperature difference between the inlet and outlet water. Surrounding liquid LPG must absorb latent heat from flowing water to evaporate. Because liquid LPG absorbs latent heat from flowing water at relatively low flow rates, it cools faster than water at relatively high flow rates. Using Equation (1), however, the heat transfer rate is greater for relatively high water flow rates. As a result, for relatively high flow rates, the surrounded liquid LPG evaporates in less time.

5. Conclusions

The results of the study demonstrate that liquid LPG can be used to instantly lower water temperature because liquid LPG has the potential to produce refrigerating effects under ambient conditions before evaporating. When water was flowing through the coil at a low rate, the temperature dropped significantly more than when it was flowing at a high rate, but high flow rates had a higher rate of heat transfer, which caused surrounding liquid LPG to evaporate more quickly than it did at low flow rates.

Author Contributions: Conceptualization, A.M. and T.A.C.; methodology, M.H.T. and M.T.; software, A.M. and A.A.; validation, A.M. and T.A.C. formal analysis, A.M.; investigation, M.H.T. and M.T.; writing—original draft preparation, A.M.; writing—review and editing, A.M. and A.A.; supervision, T.A.C. All authors have read and agreed to the published version of the manuscript.

Funding: This Research received no external funding.

Institutional Review Board Statement: Not applicable.

Informed Consent Statement: Not applicable.

Data Availability Statement: Not applicable.

Acknowledgments: The authors acknowledge the technical and financial support of the GIK Institute of Engineering Sciences and Technology, Topi 23460, Pakistan.

Conflicts of Interest: The authors declare no conflict of interest.

References

1. Setiyo, M.; Soeparman, S.; Hamidi, N.; Wahyudi, S. Cooling Effect Characteristics of a $\frac{1}{2}$ Cycle Refrigeration System on an LPG Fuel System. *Int. J. Refrig.* **2017**, *82*, 227–237. [CrossRef]
2. Muzaffar, A.; Cheema, T.A.; Abbas, A.; Tayyab, M.; Ilyas, M.; Park, C.W. Performance Analysis of Liquified Petroleum Gas (LPG) Driven Half-Cycle Air Conditioning System. *Heat Mass Transfer* **2020**, *56*, 3177–3197. [CrossRef]
3. Palmer, T.; Hill, N.; Einmen, J.V.; Simbolotti, G.; Tosato, G. Automotive LPG and Natural Gas Engines. In *ETSAP & IEA-Technology Brief T03:1–5*; ETSAP: New York, NY, USA, 2010.
4. Sarkar, J.; Bhattacharyya, S. Assessment of Blends of CO_2 with Butane and Isobutane as Working Fluids for Heat Pump Applications. *Int. J. Therm. Sci.* **2009**, *48*, 1460–1465. [CrossRef]
5. Setiyo, M.; Soeparman, S.; Wahyudi, S.; Hamidi, N. A Simulation for Predicting Potential Cooling Effect on LPG-Fuelled Vehicles. In Proceedings of the AIP Conference Proceedings, Tomsk, Russia, 22–25 November 2016; p. 030002.

engineering proceedings

Proceeding Paper

Thermo-Fluid Performance Enhancement Using NACA Aerofoil Cross-Sectional Tubes †

Muhammad Hasnain Tariq, Farooq Khan, Hafiz Muhammad Rizwan and Taqi Ahmad Cheema *

Faculty of Mechanical Engineering, GIK Institute of Engineering Sciences and Technology, Topi 23460, Pakistan
* Correspondence: tacheema@giki.edu.pk

† Presented at the 2nd International Conference on Advances in Mechanical Engineering (ICAME-22), Islamabad, Pakistan, 25 August 2022.

Abstract: Industrial heat exchange applications encounter flow across a bank of tubes in an aligned or staggered configuration. The former arrangement causes boundary layer separation and wake formation in the trailing part of the first tube leading to poor heat exchange. Alternatively, the staggered arrangement is used for heat transfer improvement, accompanied by a rise in the pressure drop. The present study uses tubes of NACA airfoil cross-sections as an alternative solution. The pressure drop and heat transfer rates in aligned aero tubes are improved by 36% and 3% more than in the circular tubes with a staggered arrangement, respectively.

Keywords: aligned tubes; staggered arrangement; heat transfer enhancement; NACA airfoils; pressure drop

check for updates

Citation: Tariq, M.H.; Khan, F.; Rizwan, H.M.; Cheema, T.A. Thermo-Fluid Performance Enhancement Using NACA Aerofoil Cross-Sectional Tubes. *Eng. Proc.* **2022**, *23*, 13. https://doi.org/10.3390/engproc2022023013

Academic Editors: Mahabat Khan, M. Javed Hyder, Muhammad Irfan and Manzar Masud

Published: 20 September 2022

1. Introduction

Cross-external flow heat exchangers are most utilized in various industrial applications such as the generation of steam in a boiler or air cooling in an air conditioner's coil [1]. The thermo-fluid performance of these heat exchangers is regularly affected by the vortex's formation, boundary layer separation, and wake regions that are downstream of heat exchanging tubes in the bank. Moreover, the higher wake formations cause significant damage to the structural stability of the tube arrays inside the bank. Therefore, improving the overall performance of these heat exchangers is critical for increasing their energy efficiency and structural stability. El Gharbi et al. [2] studied the circular, cam-shaped, and ellipsoidal-shaped tubes from the viewpoint of heat transfer characteristics and showed that the circular-shaped tubes are worse and elliptical tubes are the best. Elmekawy et al. [3] demonstrated that attaching the splitter plates at downstream ends of tubes reduces the pressure drop and increases the Nusselt number. Xu et al. [4] and Kim et al. [5] studied the effects of different fin configurations on the heat transfer rate and pressure drop and showed that the staggered configuration is better than the aligned configuration. The previous research motivates the authors to use aero-shaped tubes in the bank for both arrangements. The present study numerically investigates and compares the thermo-fluid performance of tube banks with symmetrical NACA Aero foil and circular tubes in an aligned and staggered arrangement. To the best of the author's knowledge, there is no comparison between these tubes in the literature. The results of numerical simulations are validated by the analytical results. The effect of fluid velocity on the outlet temperature and pressure drop is investigated for all configurations. The streamline contours are plotted for wake region visualization. The purpose of the study is to sustain the aligned arrangement by altering the geometry of the circular tubes instead of going for the staggered one because of the greater pressure drop.

2. Mathematical Model for Flow across Banks of Tubes

The design specifications and operating parameters for the flow across banks of tubes (circular and aero) with aligned and staggered arrangements are given here. Normally, a fluid flows over the tubes with another fluid flowing inside at a different temperature. These tube rows are arranged either in an aligned or staggered configuration with an axis normal to flow. The 2-D geometrical representation of these arrangements is shown in Figure 1. Both aligned and staggered configurations are characterized by the transverse pitch S_T, diagonal pitches S_D, and longitudinal pitch S_L measured between the tube centers. The maximum thickness of the NACA airfoil or aero tube is equal to the circular tube diameter, while the transverse and longitudinal pitches are the same for both cases, as shown in Figure 1. The fluid enters the bank with a certain value of velocity V (range: 6–21 m/s) and inlet temperature (288 K) and moves over the tubes with a surface temperature T_s (343 K) and the phenomena of momentum and energy transport take place inside the bank. The flow conditions inside the bank are dominated by the wake interactions and the boundary layer separation effects which affect the convection heat transfer. Therefore, the rate of convection heat transfer q and pressure drop Δp are evaluated here for each configuration.

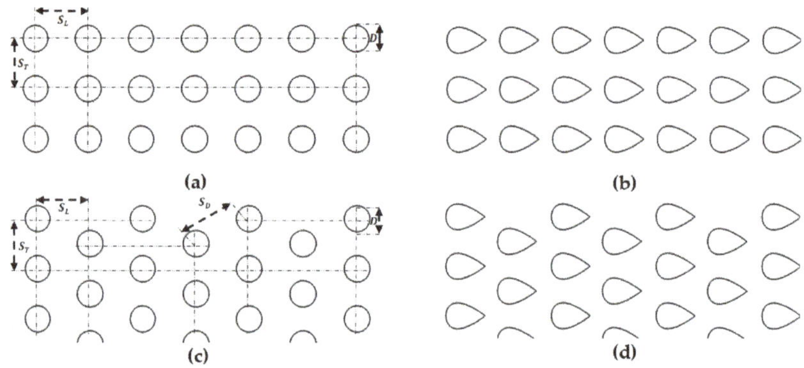

Figure 1. Tube arrangements in a bank: (**a**) circular aligned, (**b**) airfoil aligned, (**c**) circular staggered, and (**d**) airfoil staggered.

During analytical modeling, the following assumptions were made. The air is selected as a working fluid and uniform velocity is applied at the inlet of each bank. The thermophysical properties of air (ρ = 1.217 kg/m^3, c_p = 1007 J/kg-K, ν = 14.82 × 10^6 m^2/s, and k = 0.0253 W/m-K) are assumed to be constant. The fluid flow inside the bank is two-dimensional, and the heat transfer is steady state.

The Nusselt number (N_u) correlation for the fluid flow across banks of tubes in the aligned and staggered arrangement is proposed by Zukauskas [6]:

$$N_u = C_2 C {Re_{max}}^m Pr^{0.36} \left(\frac{Pr}{Pr_s}\right)^{\frac{1}{4}} \tag{1}$$

where Pr represents the Prandtl number and all properties are determined using the average inlet and outlet temperatures of fluid [1].

The Reynolds number (Re_{max}) for the aforementioned correlation depends upon the maximum fluid velocity (V_{max}) happening inside the bank and is given by Equations (2) and (3) as follows [1]:

$$Re_{max} = \frac{V_{max} D}{\nu} \tag{2}$$

$$V_{max} = \frac{S_T}{S_T - D} V \tag{3}$$

The convection heat transfer coefficient (h) and fluid outlet temperature (T_o) are determined using Equations (4) and (5) as follows [1]:

$$h = N_u \frac{k}{D} \quad (4)$$

$$T_o = T_s - (T_s - T_i)\exp\left(-\frac{\pi DNh}{\rho V N_T S_T c_p}\right) \quad (5)$$

where N represents number of tubes in the entire bank and N_T is the number of transverse plane tubes.

Log-mean temperature difference (ΔT_{lm}) and the rate of heat transfer (q') per unit length of the tubes are evaluated using Equations (6) and (7) as follows [1]:

$$\Delta T_{lm} = \frac{(T_s - T_i) - (T_s - T_o)}{ln\left(\frac{T_s - T_i}{T_s - T_o}\right)} \quad (6)$$

$$q' = N(h\pi D\Delta T_{lm}) \quad (7)$$

The pressure drop (Δp) is computed using the equation as follows:

$$\Delta p = N_L x\left(\frac{\rho {V_{max}}^2}{2}\right)f \quad (8)$$

where N_L is the number of longitudinal plane tubes, x is the correction factor, and f is the friction factor [1].

3. Results and Discussion

Figure 2a shows the variation in the weighted average temperature at the outlet as a function of the velocity of air. The maximum temperature is observed in the case of airfoil tubes in a staggered arrangement, and the minimum in the case of circular tubes in the aligned arrangement. The outlet temperature decreases by increasing the velocity because the contact time for convection heat transfer decreases inside the bank. Therefore, momentum diffusivity dominates the thermal diffusivity by increasing the velocity of fluid.

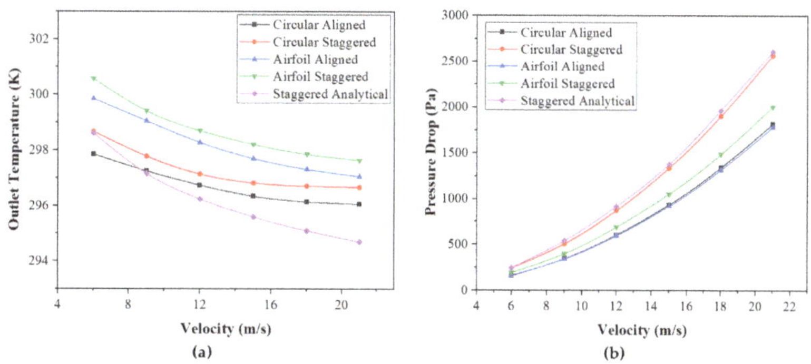

Figure 2. Effect of the fluid inlet velocity on (**a**) outlet temperature and (**b**) pressure drop.

Figure 2b depicts the variation in the pressure drop inside the bank as a function of the velocity of air. The numerical model is validated with the analytical results of circular tubes in a staggered arrangement. The pressure drop increases by increasing the velocity because the maximum fluid velocity increases at higher velocities as defined by Equation (8). The pressure drop is at a maximum in the case of circular tubes arranged in a staggered arrangement, and at a minimum for aero tubes in the aligned arrangement. Therefore, more pumping power is required for the fluid flow across the banks of circular tubes in a staggered arrangement and minimum in the case of aero tubes in the aligned arrangement.

Figure 3 shows the velocity streamlines across the banks of tubes for all configurations. The wake regions are at a maximum in the case of aligned circular tubes, and at a minimum in the case of staggered aero tubes. These boundary layer separation effects and wake interactions with the fluid flow affect the convection heat transfer inside the bank. Therefore, the net heat transfer rate is at a maximum in the case of staggered aero tubes and vice versa.

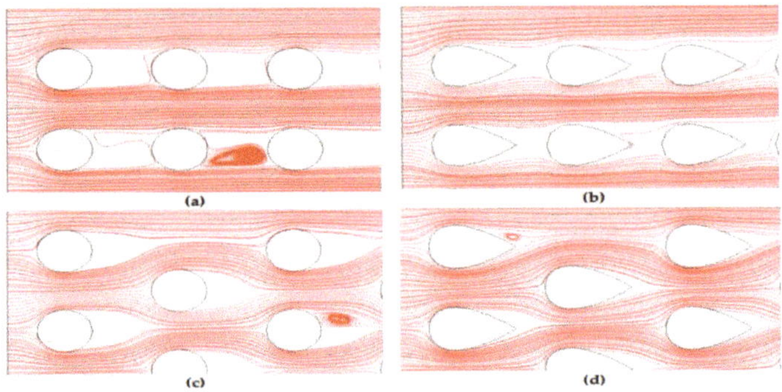

Figure 3. Streamlines around (**a**) circular aligned, (**b**) airfoil aligned, (**c**) circular staggered, and (**d**) airfoil staggered.

4. Conclusions

In the present study, the numerical results are validated with the analytical results that show a fair argument between them. The aero tubes reduce the pressure drop and increase the heat transfer rate inside the bank. The pressure drop in aligned aero tubes is 36% less than the staggering circular tubes at all velocities. Similarly, at lower velocities, the heat transfer rate is 3% greater in aligned aero tubes than in the staggering circular tubes. Therefore, at lower velocities, an aligned arrangement with aero tubes is better to use than going for circular tubes in a staggered arrangement which increases the pressure drop.

Author Contributions: Conceptualization, methodology, and analysis has been carried out by M.H.T., F.K. helped in the preparation of initial draft while H.M.R. contributed in the validation of results. Overall, this study has been completed under the supervision of T.A.C. All authors have read and agreed to the published version of the manuscript.

Funding: This research received no external funding.

Institutional Review Board Statement: Not applicable.

Informed Consent Statement: Not applicable.

Data Availability Statement: Not applicable.

Acknowledgments: The authors acknowledge the technical and financial support of GIK Institute of Engineering Sciences and Technology, 23460, Topi, Pakistan.

Conflicts of Interest: The authors declare no conflict of interest.

References

1. Incropera, F.P.; De Witt, D.P.; Bergman, T.L.; Lavine, A.S. *Introduction to Heat Transfer*, 5th ed.; Wiley: Hoboken, NJ, USA, 2002.
2. El Gharbi, N.; Kheiri, A.; El Ganaoui, M.; Blanchard, R. Numerical Optimization of Heat Exchangers with Circular and Non-Circular Shapes. *Case Stud. Therm. Eng.* **2015**, *6*, 194–203. [CrossRef]
3. Elmekawy, A.M.N.; Ibrahim, A.A.; Shahin, A.M.; Al-Ali, S.; Hassan, G.E. Performance Enhancement for Tube Bank Staggered Configuration Heat Exchanger—CFD Study. *Chem. Eng. Processing—Process Intensif.* **2021**, *164*, 108392. [CrossRef]
4. Xu, X.; Ma, T.; Li, L.; Zeng, M.; Chen, Y.; Huang, Y.; Wang, Q. Optimization of Fin Arrangement and Channel Configuration in an Airfoil Fin PCHE for Supercritical CO2 Cycle. *Appl. Therm. Eng.* **2014**, *70*, 867–875. [CrossRef]

5. Kim, T.H.; Kwon, J.G.; Yoon, S.H.; Park, H.S.; Kim, M.H.; Cha, J.E. Numerical Analysis of Air-Foil Shaped Fin Performance in Printed Circuit Heat Exchanger in a Supercritical Carbon Dioxide Power Cycle. *Nucl. Eng. Des.* **2015**, *288*, 110–118. [CrossRef]
6. Žukauskas, A. Heat Transfer from Tubes in Crossflow. *Adv. Heat. Transf.* **1972**, *8*, 93–160. [CrossRef]

engineering proceedings

MDPI

Proceeding Paper

Thermal Stability Analysis of a PCM-Based Energy Storage System †

Muhammad Umar Munir * and Abid Hussain

Department of Mechanical Engineering, Faculty of Mechanical & Aeronautical Engineering, University of Engineering and Technology Taxila, Taxila 47050, Punjab, Pakistan
* Correspondence: umarkalroo@gmail.com
† Presented at the 2nd International Conference on Advances in Mechanical Engineering (ICAME-22), Islamabad, Pakistan, 25 August 2022.

Abstract: A key factor for increasing the life of electronic devices and preventing early failure is the implementation of thermal management techniques. In thermal management, phase change materials (PCMs) are generally used. There have been a few studies conducted on PCM stability. Using thermal cycle tests, PCM (RT 42)-based energy storage systems with and without pin fins were evaluated for thermal stability. The material used for the pin fins and the heat sink was Aluminum 2024-T851. During the thermal cycle tests, the PCM-based heat sinks at 10 W had a maximum temperature difference of 1.088 °C. This leads to the PCM-based heat sink being stable during charging and discharging. According to the results of the thermal cycle tests conducted on the PCM and triangular pin-fin-based heat sink, the maximum temperature difference between the tests was 0.58 °C at 10 W. Based on the results, the PCM triangular pin-fin heat sink is stable during charging and discharging.

Keywords: thermal cycle tests; thermal stability; phase change material; thermal charging and discharging

Citation: Munir, M.U.; Hussain, A. Thermal Stability Analysis of a PCM-Based Energy Storage System. *Eng. Proc.* **2022**, *23*, 20. https://doi.org/10.3390/engproc2022023020

Academic Editors: Mahabat Khan, M. Javed Hyder, Muhammad Irfan and Manzar Masud

Published: 21 September 2022

1. Introduction

Sharma et al. [1] evaluated whether urea was conducted, and the thermal cycle tests showed a change in the latent heat of the fusion and melting point of −21% and −23.6 °C, respectively. It was observed that the urea did not melt after 50 cycles and −21%. Tauseef-ur-Rehman et al. [2] investigated the operating time behavior of an unfinned heat sink cavity with PCM. Sodhi et al. [3], depending on the criteria, described that the system might have advantages over mono PCM systems. Compared to multi-PCMs, a single PCM system can charge and discharge more efficiently. Thermal optimization using phase change materials-based systems has been refined to prevent premature equipment malfunction and maintain equipment reliability by Atouei et al. and Ali [4,5]. PCMs with high latent heat storage capability, excellent thermal stability, and adequate melting/freezing temperatures have sparked an interest in thermal control applications such as thermal management electronic devices, as evidenced by Motahar et al. [6].

2. Experimental Setup

An experimental setup is illustrated in Figure 1, with labels for all of the components used in the experiment. A PCM (RT-42) was used for the experiment. In recent experimental research investigations, A. Arshad [7] reported that in the passive thermal management of portable electronic devices, a heat sink that occupies 9% of the volume fraction is the most efficient, given as:

$$\psi_{\text{PCM}} = [\frac{V_{\text{PCM}}}{V_s - V_f}] \tag{1}$$

In Equation (1), the PCM volume fraction is denoted by ψ_{PCM}, the heat-sink volume by V_s, and the fin volume by V_f. This experiment used 220 mL or 0.22 kg PCM (RT-42) with a power supply of 10 W.

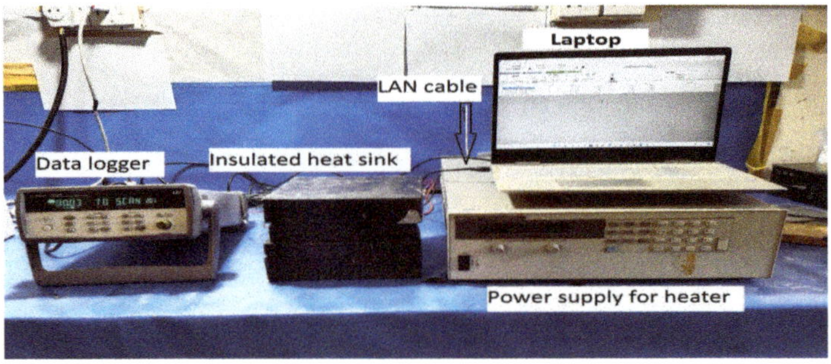

Figure 1. Actual view of the experimental setup.

Thermocouples Position

In that experiment, a K-type thermocouple was used. Figure 2 shows that the four thermocouples (T_1, T_2, T_3, and T_4) were positioned at various heights throughout the cavity.

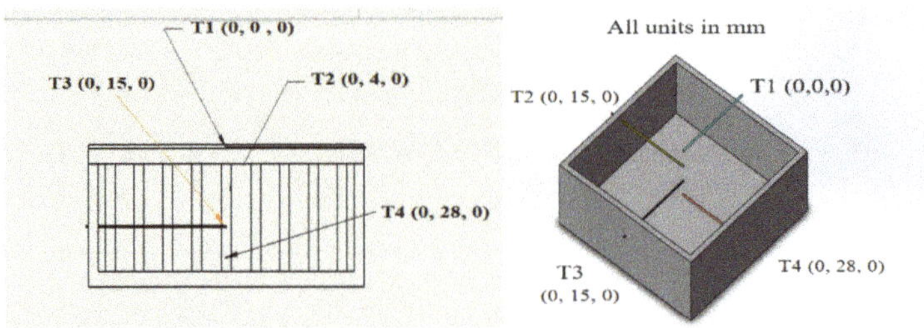

Figure 2. Schematic of thermocouple position. Left, 2D Front View of thermocouples position. Right, 3D Thermocouple Coordinate in Heat Sink.

3. Results and Discussion

3.1. Experimentally Calibration of the Heat Sink

This experiment model is validated with a simple cavity. A comparison of the experimental results of the heat sink and experimental cases can be found in Figure 3. The experimental results supported the validity of the current experimental model, indicating that the current model could be used in future studies.

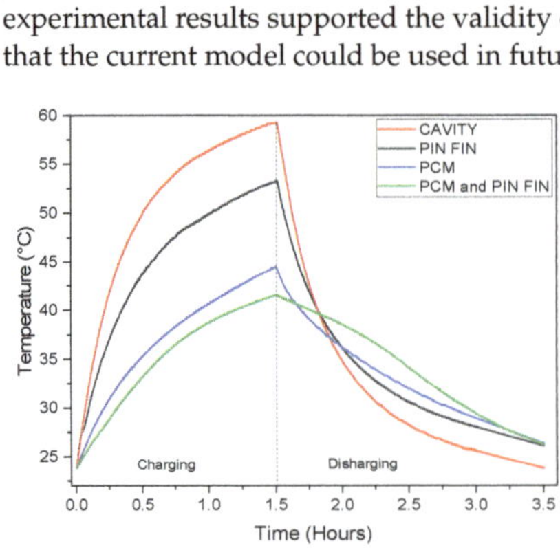

Figure 3. Experimental Validation for the heat sink.

3.2. Analysis of Thermal Stability of PCM-Based Energy Storage System

A temperature versus time diagram is depicted in Figure 4, detailing the charging and discharging for different thermal cycle tests on PCM-based heat sinks. This experiment conducted twenty thermal cycling tests to determine the stability. The thermal cycle tests were stopped because the temperature rarely changed during charging and discharging. To calculate the results and compare the thermal cycles, 05, 10, 15, and 20 were selected. Figure 4 shows the comparison of the thermocouple 4 results. After charging for 1.5 h, the maximum temperature for thermal cycle 20 was 44.821 °C, and the minimum temperature for thermal cycle 05 was 43.733 °C; so, there was a maximum temperature difference of 1.088 °C, as shown in Table 1, and all of the thermal cycles reached room temperature after discharging for 3.25 h, as shown in Figure 5.

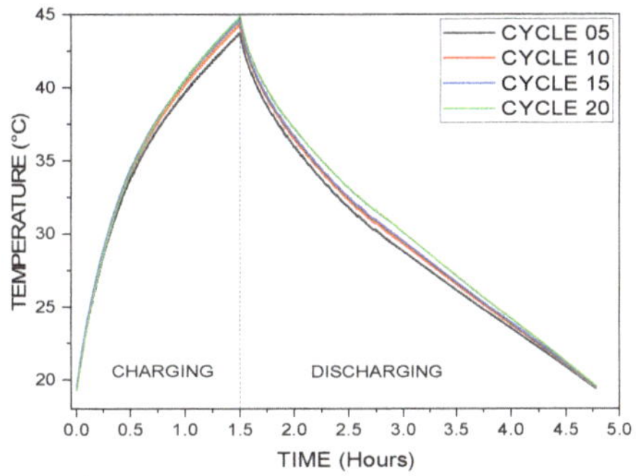

Figure 4. Thermal cycle tests analysis of PCM-based energy storage system.

Table 1. After charging, the highest temperature, and the temperature differential for Thermal Cycle Tests (TCTs).

TCTs FOR PCM-BASED HEAT SINK MAXIMUM TEMPERATURE				TCTs FOR PCM & PIN FIN BASED HEAT SINK MAXIMUM TEMPERATURE			
Cycle 5 (C_5)	**Cycle 10 (C_{10})**	**Cycle 15 (C_{15})**	**Cycle 20 (C_{20})**	**Cycle 5 (C_5)**	**Cycle 10 (C_{10})**	**Cycle 15 (C_{15})**	**Cycle 20 (C_{20})**
43.733 °C	**44.286 °C**	**44.582 °C**	**44.821 °C**	**39.429 °C**	**38.884 °C**	**39.139 °C**	**38.849 °C**
DIFFERENCE IN TEMPERATURE DURING TCTs				**DIFFERENCE IN TEMPERATURE DURING TCTs**			
	C_{10}–C_5	0.553 °C			C_5–C_{10}	0.545 °C	
	C_{15}–C_5	0.849 °C			C_5–C_{15}	0.29 °C	
	C_{20}–C_5	**1.088 °C**			**C_5–C_{20}**	**0.58 °C**	
	C_{15}–C_{10}	0.296 °C			C_{15}–C_{10}	0.255 °C	
	C_{20}–C_{10}	0.535 °C			C_{10}–C_{20}	0.035 °C	
	C_{20}–C_{15}	0.239 °C			C_{15}–C_{20}	0.29 °C	

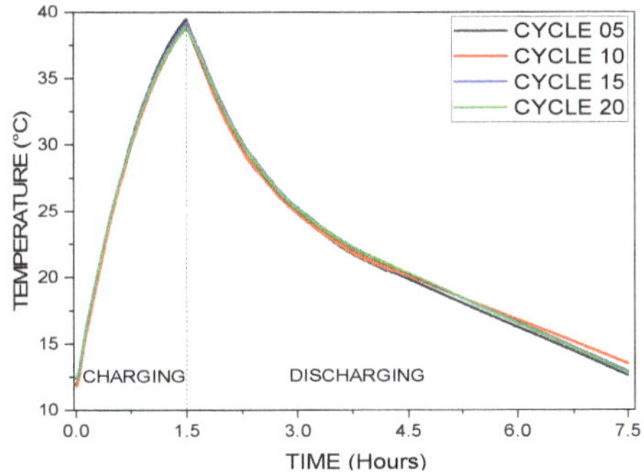

Figure 5. Thermal cycle tests analysis of PCM and triangular pin fin-based energy storage system.

processing of the model are obtained, and the two are compared so that the correctness of the similarity function relationship can be verified. The proposed research work aims to establish an accurate relationship between the actual-size model and the similar-size model based on the similar function relationship.

2. Research Methodology

2.1. CFD Foundation

Methods for studying flow problems generally include traditional experimental measurement methods, theoretical analysis methods, and CFD numerical simulation methods. The above three methods can be regarded as a complete set of research method systems. The results obtained by the experimental test method are the basis for the other two methods. However, the construction and measurement of experimental systems are often limited by various practical operations. In the current research study, AnsysFluent simulation software is used to import the fluid flow file. Firstly, single-tube models of actual size and similar size are established. Then, the two single-tube models are used to simulate simple working conditions, and the obtained temperature field distribution is completely consistent, which verifies the correctness of the derived similar function relationship.

2.2. Establishment of a Single-Tube Model

The parameters of the actual-size model and similar-scale-based model are given in Table 1. This paper determines the similar-scale factor as 20 [10]. The surface boundary is set to a constant temperature boundary, and the hourly temperature change data are imported through a user-defined function (UDF).

Table 1. Parameter settings in single-tube model.

Model Parameter	Unit	Actual Size	Similar Size
Storage length	m	20	1
Storage width	m	8	0.4
Buried tube length	m	16	0.8
Inlet water temperature	°C	40	40
Inlet flow rate	m/s	0.5	10
Radial far boundary	-	18	18
Surface boundary	-	air temperature	air temperature
Step size	S	10	10
Step count	-	25,920	65
Total length	h	72	0.18

3. Results

Solving Single-Tube Model

The similar-size model and actual model are solved according to the above model size design and operation parameter settings. After the solution, of the obtained results of both models are analyzed from the temperature field distribution and the temperature curve distribution. The cloud map of the temperature field distribution is shown in Figure 1. It can be seen from the temperature field distribution diagram that although the calculated number of steps and duration are different for the two models, the obtained temperature field distribution is basically the same. In order to more clearly compare the temperature fields of the two models, this paper takes a line at the depth of the middle of the model, which is located in the middle of the model and runs horizontally through the model. The temperature data on the line are taken out, and the temperature distribution curve can be drawn, as shown in Figure 2.

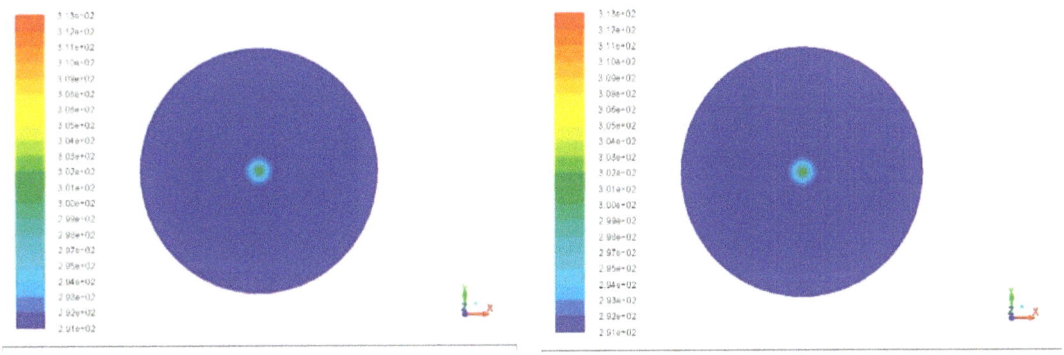

(**a**) (**b**)

Figure 1. Comparison of temperature field distribution (**a**,**b**) in single-tube model.

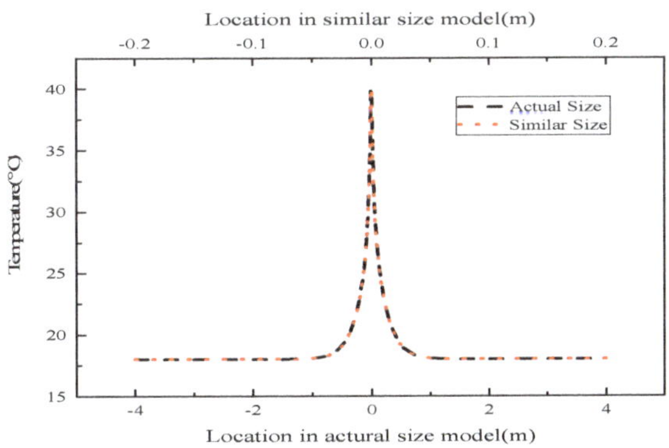

Figure 2. Comparison of temperature curve distribution in single-tube model.

In Figure 2, the horizontal axis at the top represents the position coordinates in the similar-scale model, and the corresponding red dotted line is the temperature distribution; the horizontal axis at the bottom shows the position coordinates in the actual-size model, the corresponding black dotted line is the temperature distribution in the actual-size model. It can be seen from the figure that although the two models are different in spatial scale and the calculated steps and duration are different, the soil temperature distribution calculated by the two models is the same, which fully proves the accuracy of the proposed data.

4. Conclusions

In this research work, the CFD traditional model and similar model of a soil thermal storage system are established for a single tube, and the analysis and verification are carried out step by step. The specific content and conclusions are: Firstly, single-tube models of actual size and similar size are established. Then, the two single-tube models are used to simulate simple working conditions, and the obtained temperature field distribution is completely consistent, which verifies the correctness of the derived similar function relationship. It can be considered that the proposed modeling method can be implemented for a tube group model using complex conditions, and can save a large amount of required computational power.

Author Contributions: Conceptualization, Z.A.; methodology, Z.A. and S.A.; software, Z.A. and D.C.; validation, Z.A., S.A., and M.W.K.; writing—original draft preparation, Z.A. and M.W.K.; writing—review and editing, Z.A., S.A., D.C., and M.W.K. All authors have read and agreed to the published version of the manuscript.

Funding: This research received no external funding.

Institutional Review Board Statement: Not applicable.

Informed Consent Statement: Not applicable.

Data Availability Statement: Not applicable.

Acknowledgments: The authors are thankful to Institute of Refrigeration and Cryogenics, Shanghai Jiao Tong University, China and NFC IET Multan, Pakistan for providing the opportunity to finish this research work.

Conflicts of Interest: The authors declare no conflict of interest.

References

1. Mazzarella, L.; Pedrocchi, E. Monitoring a Solar-Assisted Heat Pump with Seasonal Ground Coupled Storage: Analysis of the First Results. In Proceedings of the First European Symposium on Air Conditioning and Refrigeration, Brussels, Belgium, 5–6 November 1986; pp. 199–209.
2. van den Brink, G.J.; Hoogendoorn, C.J. Ground Water Flow Heat Losses for Seasonal Heat Storage in the Soil. *Sol. Energy* **1983**, *30*, 367–371. [CrossRef]
3. Peltola, S.S.; Lund, P.D. Comparison of Analytical and Numerical Modeling Approaches for Sizing of Seasonal Storage Solar Heating Systems. *Sol. Energy* **1992**, *48*, 267–273. [CrossRef]
4. Yumrutas, R.; Kaska, Ö. Experimental Investigation of Thermal Performance of a Solar Assisted Heat Pump System with an Energy Storage. *Int. J. Energy Res.* **2004**, *28*, 163–175. [CrossRef]
5. Inalli, M.; Ünsal, M.; Tanyildizi, V. A Computational Model of a Domestic Solar Heating System with Underground Spherical Thermal Storage. *Energy* **1997**, *22*, 1163–1172. [CrossRef]
6. Beier, R.A.; Smith, M.D.; Spitler, J.D. Reference Data Sets for Vertical Borehole Ground Heat Exchanger Models and Thermal Response Test Analysis. *Geothermics* **2011**, *40*, 79–85. [CrossRef]
7. Sakellariou, E.; Axaopoulos, P.; Wright, A. Energy and Economic Evaluation of a Solar Assisted Ground Source Heat Pump System for a North Mediterranean City. *Energy Build* **2020**, *231*, 110640. [CrossRef]
8. Ratchawang, S.; Chotpantarat, S.; Chokchai, S.; Takashima, I.; Uchida, Y.; Charusiri, P. A Review of Ground Source Heat Pump Application for Space Cooling in Southeast Asia. *Energies* **2022**, *15*, 4992. [CrossRef]
9. Abbas, Z.; Chen, D.; Li, Y.; Yong, L.; Wang, R.Z. Experimental Investigation of Underground Seasonal Cold Energy Storage Using Borehole Heat Exchangers Based on Laboratory Scale Sandbox. *Geothermics* **2020**, *87*, 101837. [CrossRef]
10. Abbas, Z.; Yong, L.; Abbas, S.; Chen, D.; Li, Y.; Wang, R.Z. Performance Analysis of Seasonal Soil Heat Storage System Based on Numerical Simulation and Experimental Investigation. *Renew Energy* **2021**, *178*, 66–78. [CrossRef]

engineering proceedings

Proceeding Paper

Hydrodynamic and Thermal Energy Characteristics of a Gravitational Water Vortex †

Hafiz Muhammad Rizwan, Taqi Ahmad Cheema *, Muhammad Hasnain Tariq and Atif Muzaffar

Faculty of Mechanical Engineering, GIK Institute of Engineering Sciences and Technology, Topi 23460, Pakistan
* Correspondence: tacheema@giki.edu.pk
† Presented at the 2nd International Conference on Advances in Mechanical Engineering (ICAME-22), Islamabad, Pakistan, 25 August 2022.

Abstract: In the present study, a numerical analysis has been conducted to investigate the hydrodynamic and thermal energy transfer capacity of a vortex formed under the effect of gravity. For this purpose, the study uses a gravitational water vortex heat exchanger (GWVHE), which includes baffles around a cylindrical basin in which a water vortex is formed under the effect of gravity. The results have been examined for different inlet boundary conditions based on flow and temperature to determine the strength of vortex formation and comparative energy transfer rate for both fluid domains. A strong vortex is formed in the basin at a height to diameter ratio between 0.41 and 0.54 with a minimum inlet mass flow rate of 0.005 kg/s, which effectively increases the energy exchange potential due to the centrifugal effect. The reasonable energy agreement has been obtained for the minimum flow rates of both fluid domains; however, the thermal energy losses are increased with the increase in the inlet mass rate of the hot domain, due to the reduction in the time of contact. The existence of an acceptable energy balance and strong vortex formation at a minimum flow rate sparks the need for a new configuration to enhance the thermal performance of GWVHE.

Keywords: strong vortex formation; basin with baffles; energy balance; heat transfer rate; gravitational water vortex heat exchangers

Citation: Rizwan, H.M.; Cheema, T.A.; Tariq, M.H.; Muzaffar, A. Hydrodynamic and Thermal Energy Characteristics of a Gravitational Water Vortex. *Eng. Proc.* **2022**, *23*, 19. https://doi.org/10.3390/engproc2022023019

Academic Editors: Mahabat Khan, M. Javed Hyder, Muhammad Irfan and Manzar Masud

Published: 20 September 2022

1. Introduction

Heat exchangers (HEs) are the devices that are used in the heat transfer process without physically mixing the fluid streams. The classification of HEs is usually based on the flow arrangement or the design configuration. Parallel flow, counter flow, and cross flow HEs are the examples of flow arrangement-based HEs, whereas compact types, shell and tube HEs are classified according to their design configuration [1]. The direction of two fluid streams and the structural configuration of HEs are very important features when improving the heat exchange process. Active and passive approaches are used to reduce the thermal losses, as well as to improve the heat transfer performance in various thermal systems [2].

Another important flow configuration in HEs is a vortex flow. The fluid circulates around a central axis in a vortex flow, forming either a forced vortex or a free vortex. The free vortex is generated naturally, whereas the forced vortex is formed because of solid body rotation. The free vortex can be artificially created if a viscous fluid flows through a hole at the bottom of a tank under the force of gravity [3,4]. The present study at first determines the strength of a strong vortex under the effect of gravity called a gravitational vortex and then investigates its thermal energy exchange capacity using a GWVHE [5]. A new design of GVHE with baffles on the outer side of the cylinder and a height to diameter ratio of 0.54 has been employed to numerically investigate the energy balance by varying the inlet conditions.

2. Mathematical Modeling and Governing Equation

The 3D computational model of GWVHE used in the present study is shown in Figure 1. The height and diameter of the basin are 380 mm and 700 mm, respectively. There are two fluid domains separated by a solid domain. In both fluid domains, the fluid is water. The outer cylinder with baffles contains hot water, while the central basin cylinder has cold water. The solid domain is assumed to be made of steel. To effectively induce the gravitational effect, the top interface of the basin cylinder remains open to the atmosphere. Steady state heat transfer analysis of the GWVHE is performed using COMSOL-Multiphysics (V 6.0) and a user control mesh is applied on each domain to discretize the geometric model.

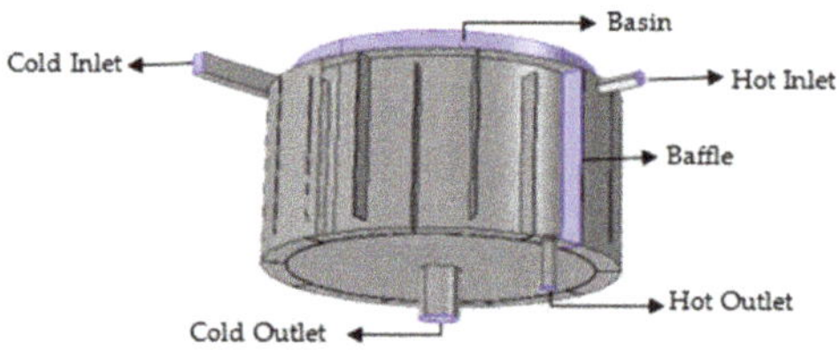

Figure 1. The 3D computational model of GWVHE.

2.1. Fluid Flow Modeling

A high Reynolds number fluid is required for the vortex formation and this is why a turbulent flow is considered in both domains. For the investigation of fluid characteristics, the turbulent k-ε model is applied on both fluid domains, along with conservation equations related to mass and momentum. The governing equations for the turbulent k-ε model are given as follows [2]:

$$\nabla\cdot\left[\left(\mu+\frac{\mu_T}{\sigma_k}\right)\nabla k\right]+\rho_l\mu_l\cdot\nabla k=\frac{1}{2}\mu_T\left(\nabla\mu_l+(\nabla\mu_l)^T\right)^2-\rho_l\varepsilon+S_k \tag{1}$$

$$\nabla\cdot\left[\left(\mu+\frac{\mu_T}{\sigma_\varepsilon}\right)\nabla\varepsilon\right]+\rho_l\mu_l\cdot\nabla\varepsilon=\frac{1}{2}C_{\varepsilon 1}\frac{\varepsilon}{k}\mu_T\left(\nabla\mu_l+(\nabla\mu_l)^T\right)^2-\rho_l C_{\varepsilon 2}\frac{\varepsilon^2}{k}+\frac{\varepsilon}{k}C_\varepsilon S_k \tag{2}$$

The values for the constants are taken as σ_k = 1.0, σ_ε = 1.3, $C_{\varepsilon 1}$ = 1.44, $C_{\varepsilon 2}$ = 1.92 and C_μ = 0.09.

2.2. Heat Transfer Modeling

The energy equations for fluid and solid domains are given as the following equations, respectively [2].

$$\rho C_{pf}\left(V_r\frac{\partial T_f}{\partial r}+\frac{V_\theta}{r}\frac{\partial T_f}{\partial\theta}+V_z\frac{\partial T_f}{\partial z}\right)=k_f\left(\frac{1}{r}\frac{\partial}{\partial r}\left(r\frac{\partial T_f}{\partial r}\right)+\frac{1}{r^2}\frac{\partial^2 T_f}{\partial\theta^2}+\frac{\partial^2 T_f}{\partial z^2}\right) \tag{3}$$

$$k_s\left(\frac{1}{r}\frac{\partial}{\partial r}\left(r\frac{\partial T_s}{\partial r}\right)+\frac{1}{r^2}\frac{\partial^2 T_s}{\partial\theta^2}+\frac{\partial^2 T_s}{\partial z^2}\right)=0 \tag{4}$$

where T_f, k_f, and C_{pf} denote the temperature, thermal conductivity, and specific heat, respectively.

3. Results and Discussion

Figure 2a–d show the velocity distributions in the form of contours that make the air core at the mid-section from top to bottom, indicating the formation of a strong vortex. The strong vortex formation in the basin is achieved at all inlet mass flow rates of the cold domain from 0.005 to 0.5 kg/s and also the ratio of height to diameter is calculated, ranging from 0.41 to 0.54 for the strong vortex formation. The strength of the gravitational vortex enhances the heat transfer effect due to the centrifugal phenomena.

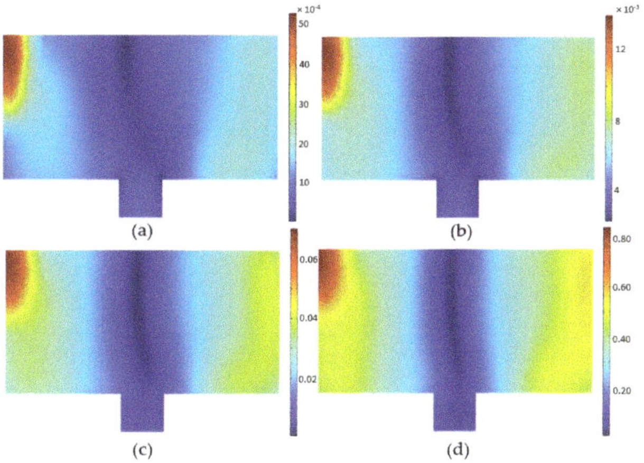

Figure 2. Formation of velocity vortex air core in the center of basin at $\dot{m}_h$ = 2.29 kg/s, (**a**) $\dot{m}_c$ = 0.005 kg/s, (**b**) $\dot{m}_c$ = 0.01 kg/s, (**c**) $\dot{m}_c$ = 0.05 kg/s, (**d**) $\dot{m}_c$ = 0.5 kg/s.

Figure 3a–d show the temperature distributions of the cold fluid domain at constant inlet mass flow of the hot domain, which represent the temperature gradient at the mid-section of the basin. It is observed from the simulation analysis that the highest temperature gradient is obtained at the lowest inlet mass flow rate of 0.005 kg/s. This shows that the time of contact between both fluids is very important to exchange energy.

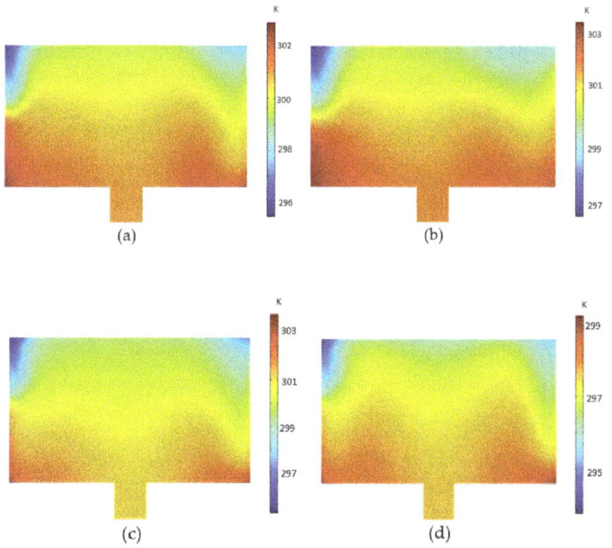

Figure 3. Temperature gradient in the center of basin at $\dot{m}_h$ = 2.29 kg/s, (**a**) $\dot{m}_c$ = 0.005 kg/s, (**b**) $\dot{m}_c$ = 0.01 kg/s, (**c**) $\dot{m}_c$ = 0.05 kg/s, (**d**) $\dot{m}_c$ = 0.5 kg/s.

Figure 4a,b show the energy balance comparison of both fluid streams with (0.005 kg/s to 0.5 kg/s) the inlet mass flow rate of the cold domain. However, the inlet mass flow of the hot domain varies from 0.002 kg/s to 2.29 kg/s. It can be observed that at the smaller flow rates of both domains, the thermal losses are less but as the hot side inlet mass flow rate increases, the energy losses to the environment are increased.

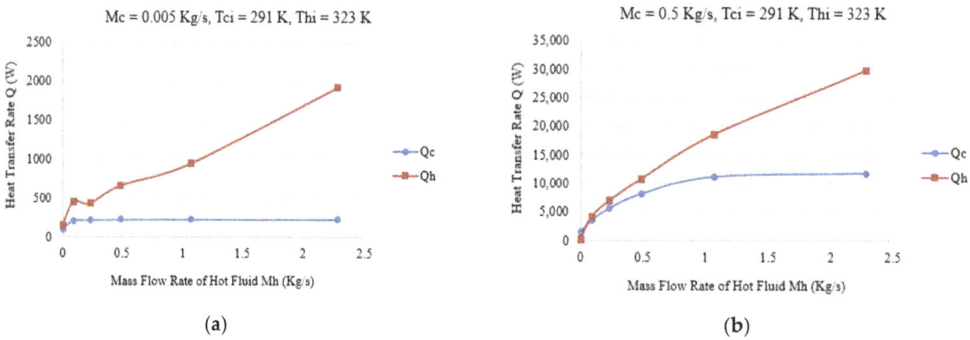

Figure 4. Energy balance comparision with different mass flow rates.

Therefore, the hot inlet mass flow rate is critical to investigate the rate of heat transfer. For the energy balance calculation, the general relationship for heat transfer rate is used for both domains.

4. Conclusions

In the present study, a simulation analysis has been performed for the investigation of GWVHE with vortex flow-based configuration. Different flow and temperature conditions are analyzed, and the computational results reveal that the purposeful configuration can be used to enhance the thermal performance of GWVHE, due to strong vortex formation at the minimum mass flow rate.

Author Contributions: Conceptualization, H.M.R. and T.A.C.; methodology, H.M.R. and A.M.; software, H.M.R. and M.H.T.; validation, H.M.R. and A.M.; formal analysis, H.M.R. and M.H.T.; investigation, H.M.R. and M.H.T.; data curation, H.M.R.; writing—original draft preparation, H.M.R.; writing—review and editing, H.M.R. and M.H.T.; supervision, T.A.C.; project administration, T.A.C. All authors have read and agreed to the published version of the manuscript.

Funding: This research received no external funding.

Institutional Review Board Statement: Not applicable.

Informed Consent Statement: Not applicable.

Data Availability Statement: Not applicable.

Acknowledgments: The authors acknowledge the support of the Interdisciplinary Engineering, Modelling and Simulation Research Group (IEMSRG) of the GIK Institute, Topi, 23460, Pakistan.

Conflicts of Interest: The authors declare no conflict of interest.

References

1. Incropera, F.P.; De Witt, D.P.; Bergman, T.L.; Lavine, A.S. *Introduction to Heat Transfer*, 5th ed.; Wiley: Hoboken, NJ, USA, 2002.
2. Tayyab, M.; Cheema, T.A.; Malik, M.S.; Muzaffar, A.; Sajid, M.B.; Park, C.W. Investigation of Thermal Energy Exchange Potential of a Gravitational Water Vortex. *Renew Energy* **2020**, *162*, 1380–1398. [CrossRef]
3. Ullah, R.; Cheema, T.A.; Saleem, A.S.; Ahmad, S.M.; Chattha, J.A.; Park, C.W. Preliminary Experimental Study on Multi-Stage Gravitational Water Vortex Turbine in a Conical Basin. *Renew Energy* **2020**, *145*, 2516–2529. [CrossRef]
4. Sachchidanand, M.; Nimankar, J.; Sachin, P.; Dahake, K. Global journal of engineering science and researches review of heat exchangers. *Nimankar* **2016**, *3*. [CrossRef]
5. Çengel Yunus, A.; Ghajar, A.J. *Heat and Mass Transfer: Fundamentals & Applications*, 4th ed.; McGraw-Hill: New York, NY, USA, 2011.

engineering proceedings

MDPI

Proceeding Paper

Harnessing Ocean Thermal Energy from Offshore Locations in Pakistan Using an Organic Rankine Cycle †

Muhammad Haroon [1,*], Abubakr Ayub [2], Nadeem Ahmed Sheikh [3], Muhammad Ahmed [1] and Al-Bara Shalaby [3]

[1] Department of Mechanical Engineering, Capital University of Science and Technology (CUST), Islamabad 44000, Pakistan
[2] Department of Mechanical and Industrial Engineering, University Degli Studi di Brescia, 25121 Brescia, Italy
[3] Department of Mechanical Engineering, International Islamic University, Islamabad 44000, Pakistan
* Correspondence: muhammad.haroon@cust.edu.pk

† Presented at the 2nd International Conference on Advances in Mechanical Engineering (ICAME-22), Islamabad, Pakistan, 25 August 2022.

Abstract: The temperature gradient of the top and bottom layers of sea water can be employed for power generation through ocean thermal energy conversion cycles (OTEC). In this thermodynamic study, an organic Rankine cycle is used to convert the ocean thermal energy of the Arabian sea into useful electrical power. A comparative investigation is carried out between R152a and R1234yf working fluids. Sensitivity analysis is performed at varying condenser saturation temperatures and evaporator saturation temperatures. For the R152a OTEC cycle, 2.8% thermal efficiency, 47.9% exergy efficiency and 9064 W turbine work are obtained at the optimum point. Similarly, for the R1234yf working fluid in the OTEC cycle, 2.8% thermal efficiency, 46.9% exergy efficiency and 4818 W turbine work are obtained at the optimum point.

Keywords: ocean thermal energy; organic Rankine cycle; Carnot efficiency; exergy efficiency

Citation: Haroon, M.; Ayub, A.; Sheikh, N.A.; Ahmed, M.; Shalaby, A.-B. Harnessing Ocean Thermal Energy from Offshore Locations in Pakistan Using an Organic Rankine Cycle. *Eng. Proc.* **2022**, *23*, 24. https://doi.org/10.3390/engproc2022023024

Academic Editors: Mahabat Khan, M. Javed Hyder, Muhammad Irfan and Manzar Masud

Published: 21 September 2022

1. Introduction

Due to incremental global electric power demand and modern international climatic stipulations, renewable energy resources such as ocean thermal energy conversion cycles (OTEC) are gaining high interest [1]. OTEC cycles exploit the temperature gradient of the top and bottom layers of sea water for power generation. Since temperature gradients of the top and bottom layers of tropical oceans usually range from 22 °C to 26 °C [2], the highest theoretical efficiency achievable from a Carnot cycle working on these temperature gradients is approximately 5–6%.

Recently, some thermodynamic investigations on OTEC systems were conducted using an organic Rankine cycle. J.I. Yoon et al introduced a highly efficient OTEC system equipped with a motive pump and liquid-vapor ejector using R152a as a working fluid. A comparative performance analysis is performed between the modified OTEC and basic OTEC systems. Results shows that the modified OTEC cycle produced higher power output than the basic OTEC cycle. Furthermore, the modified OTEC system at optimized operating conditions yields 38% gain in cycle efficiency as compared to the basic OTEC system [3]. For improved efficiency and performance, an organic Rankine cycle (ORC) for OTEC system working with R717 was also investigated in [1,4]. Bernardoni et al. [1] numerically evaluated an OTEC system for power generation using R717 working fluid. The system efficiency of OTEC was 2.2% at warm sea water temperature of 28 °C and cold sea water temperature of 4 °C. Nevertheless, a small temperature gradient between surface (warm) and cold sea water results in a limited amount of power production from the OTEC system. Therefore, in addition to modifying cycle configuration, which possibly leads to larger investment costs, another useful approach is to utilize new organic fluids or a mixture of organic fluids with good environmental properties as working fluids in a basic OTEC cycle with the aim to enhance turbine power, first law and second law efficiency.

In this study, thermodynamic performance of OTEC systems operating with R152a and R1234yf working fluids are analyzed for Arabian Sea environmental conditions. The aim is to study the effect of properties of two different working fluids on the thermodynamic performance of an OTEC system. Sensitivity analysis is performed at varying condenser saturation temperatures and evaporator saturation temperatures for both working fluids. Performance indicators considered for this parametric study are, thermal efficiency, exergy efficiency, turbine work and overall heat transfer coefficient time area (UA) of the evaporator and condenser. Performances of both OTEC cycles are compared at optimum points and the best performing working fluid is suggested for power generation.

2. Thermodynamic Model and Calculation Method

This study explored the energetic and exergetic performance of closed loop organic Rankine cycles powered by R152a and R1234yf working fluids for an OTEC system. These pure working fluids are selected on the basis of high vaporization capacity and favorable environmental attributes (Table 1). The Peng-Robinson equation (PR-EOS) of state is used in this parametric study to compute thermo-physical properties of both working fluids. In this study, all thermal calculations by employing PR-EOS are carried out in Aspen Plus V10 simulation tool.

Table 1. Properties of water in Arabian sea [5], and important thermodynamic and environmental properties of selected working fluids.

Properties	Value	Properties	R152a	R1234yf
Surface Sea water temperature	28 °C	Latent heat of Vaporization, kJ/kg (at −10 °C)	307	163
Surface Sea water pressure	1.01 bar	GWP (IPCC AR4)	124	4
Depth Sea water temperature	10 °C	Self-ignition temperature (°C)	454	405
Depth Sea water pressure	50.66 bar	ODP	0	0
Depth of Sea water	500 m	ASHRAE safety class	A2	A2L
Mass flow rate of sea water	200 Kg/s			

As a preliminary study, design point analysis is carried out using average temperature and pressure data from the Arabian sea [5] as the heat source and heat sink data (reported in Table 1). The closed loop simple cycle configuration of the ORC is considered for power production, which is presented in Figure 1a.

A thermodynamic model and simulation of an organic Rankine cycle are performed in Aspen Plus V10 process simulation software. For sensitivity analysis, the saturation temperature of an evaporator is varied between 17–24 °C and condensation temperature in a condenser is varied within 12–16 °C. Performance parameters studied are: pump work, turbine work, heat exchangers duties, UA, enthalpy and entropy generations in each cycle component, 1st law and second law thermodynamic cycle efficiency of the cycle. At last, an optimum performance point is decided for both OTEC cycles based on maximum exergy efficiency and turbine work and minimum UA values. A T-s diagram of R152a OTEC cycle at optimum performance point is presented in Figure 1b. The value of UA is calculated for both heat exchangers (evaporator and condenser) based on the LMTD and net heat duties.

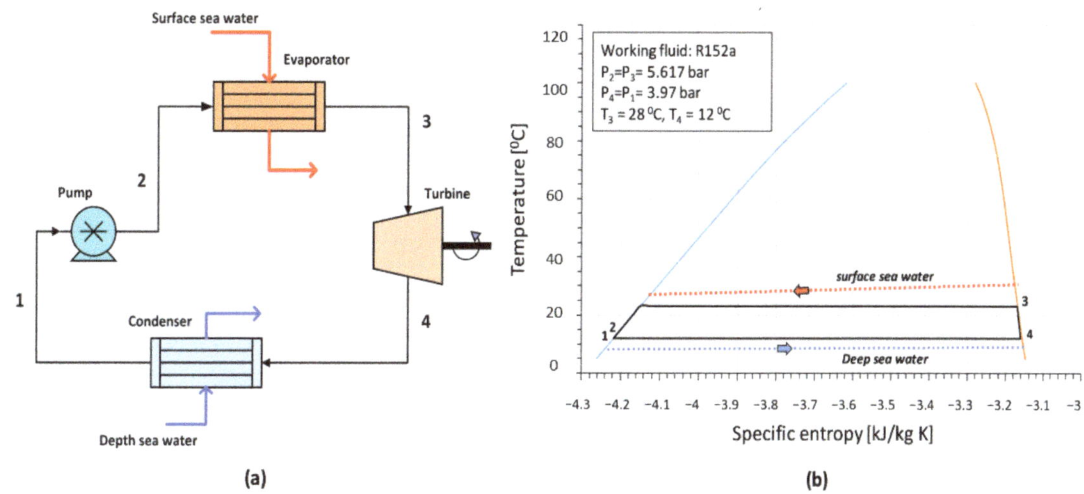

Figure 1. (**a**) Closed loop OTEC architecture along with (**b**) T-s diagram of R152a OTEC system.

Performance Parameters

The mathematical expression of first law efficiency (energy efficiency) and second law efficiency (exergy efficiency) of an organic Rankine cycle operating between warm sea water temperatures and cold sea water temperatures are defined as shown below

$$\eta_I = [Turbine\ power - Pump\ power]/Evaporator\ heat\ duty \quad (1)$$

$$\eta_{II} = \eta_I/\eta_{carnot} \quad (2)$$

Losses in cycle efficiency (η_{loss}) can be evaluated using entropy generation in each cycle component and dead state temperature (T_0), which is assumed to be equal to deep sea water temperature, i.e., 10 °C. Power input from warm sea water is the required thermal input of the ORC cycle and it is equal to the heat duty of the evaporator.

$$\eta_{loss} = \frac{T_0 \sum_{i=1}^{N} \dot{S}_G}{Evaporator\ heat\ duty} \quad (3)$$

The definitions of overall heat transfer coefficient times area of heat exchanger (UA) and log mean temperature difference (LMTD) are expressed below,

$$UA = Heat\ exchanger\ day/LMTD \quad (4)$$

$$LMTD = \frac{\Delta T_{hot,side} - \Delta T_{cold,side}}{ln(\Delta T_{hot,side}/\Delta T_{cold,side})} \quad (5)$$

3. Results and Discussion

The variation of turbine power and total UA ($UA_{evaporator} + UA_{condenser}$) with respect to variation in evaporator saturation temperature (T_{evap}) and condenser saturation temperature (T_{cond}) is demonstrated in Figure 2a,b, respectively. The increase in saturation temperature in the evaporator increases turbine inlet temperature and also turbine inlet pressure, which results in larger expansion in the turbine and larger turbine power. Similarly, by keeping the saturation temperature of the evaporator constant at 23 °C, the rise in condenser saturation temperature or the turbine outlet temperature leads to reduction in temperature ratio (T_3/T_4) across the turbine and reduction in turbine outlet pressure, and, consequently, a reduction in power output is observed, as evident from Figure 2b.

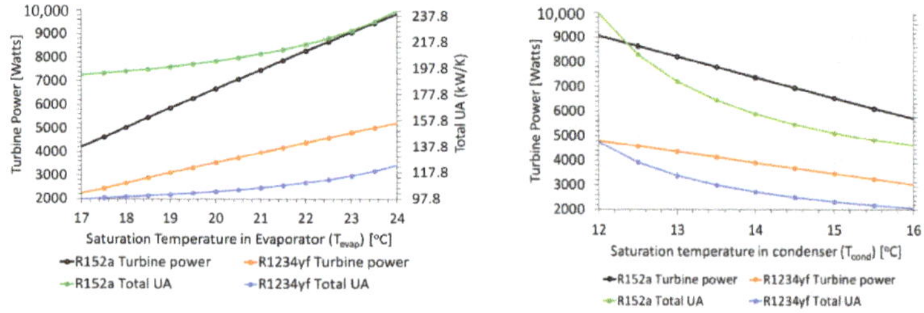

(a) (b)

Figure 2. (**a**) Variation of evaporator temperature and (**b**) Condenser temperature on turbine power and total UA of organic Rankine cycle operating with R152a and R1234yf working fluids.

In summary, higher evaporator saturation temperature (T_{evap}) and lower condenser saturation temperature (T_{cond}) results in larger power output from the turbine. Moreover, a turbine operating with R152a working fluid produced more power output compared to R1234yf working fluid at all values of evaporator saturation temperature and condenser saturation temperature. For a T_{evap} of 23 °C and T_{cond} of 12 °C, power output from R152a is 9063 W and for R1234yf, it is 4818 W. Total UA in the case of R1234yf working fluid is lower than in the case of R152a working fluid, owing to lower heat duty of the evaporator and condenser for R1234yf. For a fixed value of T_{evap}, the rise in T_{cond} brings about an increase in MITA and LMTD of the condenser, which reduces total UA.

4. Conclusions

The study explores the thermodynamic performance of organic Rankine cycles operating with R152a and R1234yf working fluid for ocean thermal energy conversion. For both working fluids, the power output increases with a rise in evaporator temperature and a reduction in condenser temperature. Secondly, the power output in the case of R152a working fluid is higher than R1234yf working fluid but the value of total UA (indicator for size of condenser and evaporator) is comparatively higher in the case of R152a working fluid. The maximum first law efficiency and second law efficiency obtained for R152a working fluid are 2.8% and 47.9%, respectively.

Author Contributions: Conceptualization, M.H. and A.A.; methodology, A.A.; software, M.H.; validation, M.A. and A.-B.S.; formal analysis, M.H.; investigation, M.H.; resources, A.A.; data curation, M.A. and A.-B.S.; writing—original draft preparation, M.H.; writing—review and editing, M.H. and A.A.; supervision, N.A.S. and A.A. All authors have read and agreed to the published version of the manuscript.

Funding: This research received no external funding.

Institutional Review Board Statement: Not applicable.

Informed Consent Statement: Not applicable.

Data Availability Statement: Not applicable.

Conflicts of Interest: The authors declare no conflict of interest.

References

1. Bernardoni, C.; Binotti, M.; Giostri, A. Techno-Economic Analysis of Closed OTEC Cycles for Power Generation. *Renew. Energy* **2019**, *132*, 1018–1033. [CrossRef]
2. Avery, W.H.; Wu, C. *Renewable Energy from the Ocean: A Guide to OTEC*, 1st ed.; Oxford University Press: Oxford, UK, 1994; ISBN 978-0195071993.
3. Yoon, J.I.; Seol, S.H.; Son, C.H.; Jung, S.H.; Kim, Y.B.; Lee, H.S.; Kim, H.J.; Moon, J.H. Analysis of the High-Efficiency EP-OTEC Cycle Using R152a. *Renew. Energy* **2017**, *105*, 366–373. [CrossRef]

4. Yang, M.H.; Yeh, R.H. Investigation of the Potential of R717 Blends as Working Fluids in the Organic Rankine Cycle (ORC) for Ocean Thermal Energy Conversion (OTEC). *Energy* **2022**, *245*, 123317. [CrossRef]
5. Niazi, B.; Amjad, S. Harnessing Ocean Energy from Coastal and Offshore Pakistan. *Engineering* **2022**, *12*, 78. [CrossRef]

engineering proceedings

MDPI

Proceeding Paper

Development and Analysis of a Liquid Piston Stirling Engine †

Abdur Rehman Mazhar *, Hassan Zamir Khan, Muhammad Kashif Khan, Adeel Ahmed and Muhammad Hassaan Yousaf

College of Electrical & Mechanical Engineering, National University of Sciences & Technology, Islamabad 47301, Pakistan

* Correspondence: arehman.mazhar@ceme.nust.edu.pk

† Presented at the 2nd International Conference on Advances in Mechanical Engineering (ICAME-22), Islamabad, Pakistan, 25 August 2022.

Abstract: Stirling engines are a type of reciprocating external combustion engine that uses one or more pistons to perform useful work with the help of some external heat. The Fluidyne design, also known as the liquid piston Stirling engine, uses liquid as pistons contained in a cylinder that entraps a working gas. Stirling engines are low efficiency engines that can utilize waste and low-grade thermal energy to pump water or do work at a small scale. Excellent opportunities to address difficulties with energy security, water dissipation, and greenhouse gas emissions are provided by the employment of these low-grade waste heat recovery techniques. This research will cover effects of changing water levels and using a mixture of different working liquids of low heat of vaporization on engine performance. Temperature, pressure, amplitude of oscillation variation with time and pumping volume were determined with Vernier sensors and tracking software. This study confirms correlation between working liquid volume and heat of vaporization on engine performance.

Keywords: Stirling engine; pumping; sustainability; low-grade heat; Fluidyne engine; external combustion engine

Citation: Mazhar, A.R.; Khan, H.Z.; Khan, M.K.; Ahmed, A.; Yousaf, M.H. Development and Analysis of a Liquid Piston Stirling Engine. *Eng. Proc.* **2022**, *23*, 34. https://doi.org/10.3390/engproc2022023034

Academic Editors: Mahabat Khan, M. Javed Hyder, Muhammad Irfan and Manzar Masud

Published: 26 September 2022

1. Introduction

The liquid piston heat engine offers unparalleled potential to both renewable and waste-heat recovering methods because of its low manufacturing and operational costs [1]. The flexible geometry of the engine holds a fluid and a gas in place [2]. The components are inexpensive and simple to manufacture [3]. Stirling engines operate on a Stirling cycle [4,5]. The manufacturing costs of a liquid piston Stirling engine are extremely low. However, the full potential of such a simple energy conversion device is still underexplored considering that it could be used in areas with limited maintenance and availability of materials [6]. A better understanding of the engine's performance parameters is provided by this research. These are affected by fluid vaporization and volume of working liquid [3,7]. Additionally the pumping volume for each case has been measured.

2. Design

Fluidyne is divided into 4 segments which are labelled in Figure 1 with the testing apparatus shown in Figure 2. The dimensions of the model are given in Figure 3 [8]. The hot end is made of copper, the cold end with clear PVC (Polyvinyl chloride) tube and the regenerator with polyethylene foam with a pumping line attached at the hot end. Heat is supplied via a hot air gun with a 2000 W rating. Vernier temperature and pressure sensors are used for measuring values of air at cold end [9].

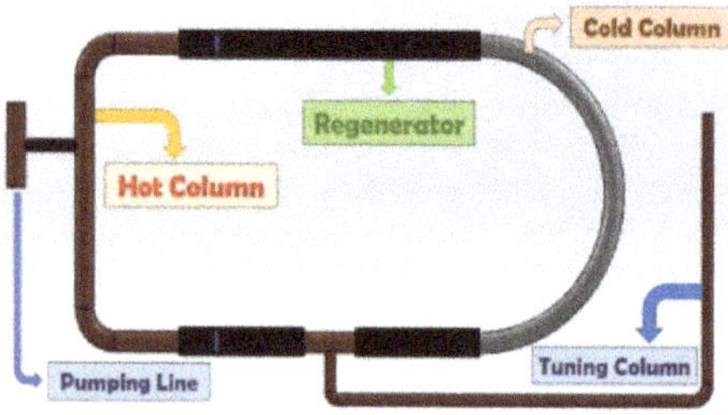

Figure 1. CAD model with segments.

Figure 2. Testing apparatus.

Section Name	Length	External Diameter	Internal Diameter
Regenerator	33 cm	1.3 cm	0.65 cm
Cold End Portion	38.1 cm	1.5 cm	1.3 cm
Hot End Portion	22.9 cm	1.3 cm	1.3 cm
Displacer Portion	33 cm	1.3 cm	1.3 cm

Figure 3. Dimensions of segments.

3. Results and Analysis

The effect of changing working liquid volume and liquid column vaporization on engine performance was analyzed. For heat input, a heat gun was utilized with a rated power of 2000 W. The amplitude of oscillation of the liquid, which is the displacement covered by the liquid from its initial to its final position in the tuning column, are measured using tracker software.

3.1. Effect of Working Liquid Volume (Water Is Used Only)

In order to analyze how an engine's working liquid volume affects performance, tests were carried out with volume of the working liquid (water) varied as 100 mL and 80 mL. The effects of the varied volumes were studied with respect to oscillations amplitude and stability, pressure and temperature variations in the cold end, and the pumping power at that volume was found. Vernier temperature and pressure sensors were used to record these values, which are illustrated in Figures 4 and 5.

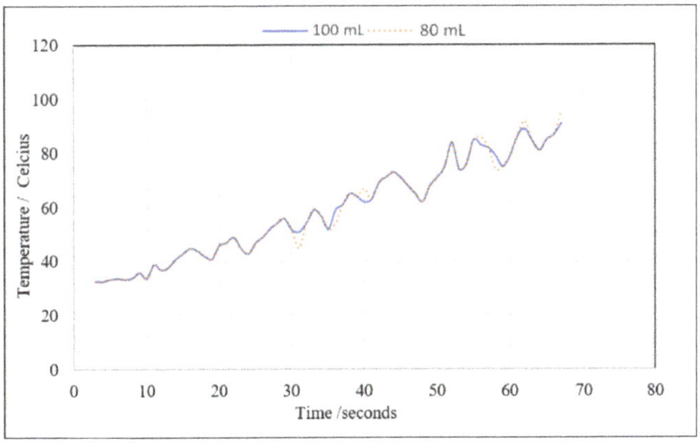

Figure 4. Temperature vs. time curve.

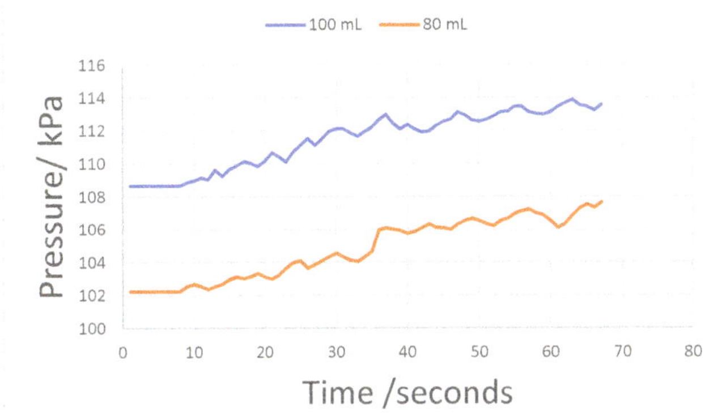

Figure 5. Pressure vs. time curve.

The oscillations produced in the tuning column were analyzed for both volumes using tracker 6.0.10 software, and the variation of amplitude with time is illustrated in Figures 6 and 7.

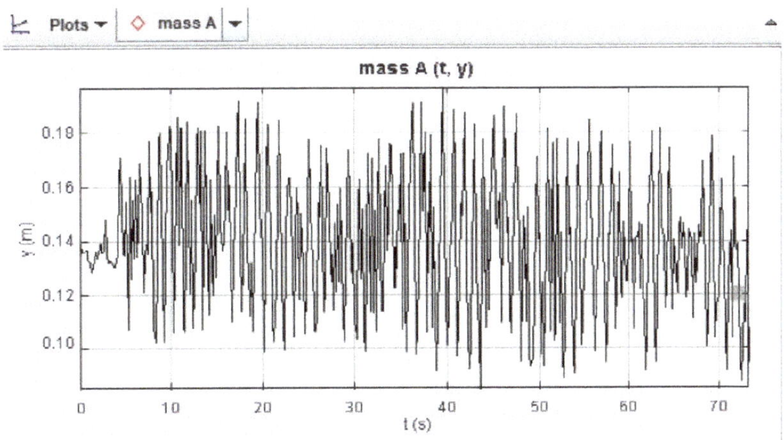

Figure 6. 80 mL volume.

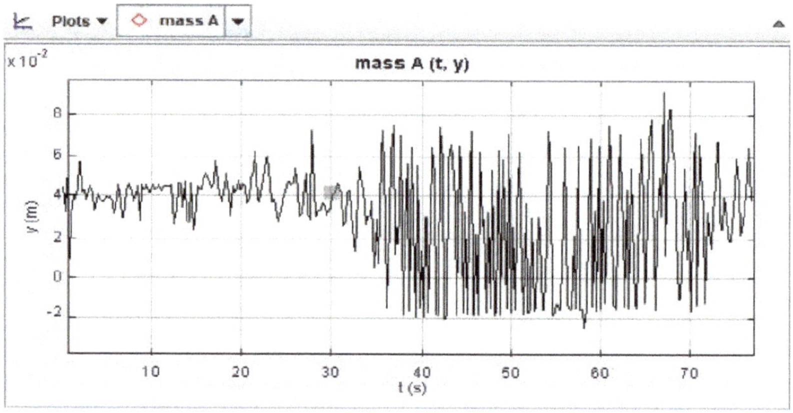

Figure 7. 100 mL volume.

To determine the pumping rate, we added a water reservoir under the pumping line, and a test tube of 20 mL capacity was connected to the upper part of the pumping line. The time required to fill the test tube was measured with a stopwatch. The time was measured from the instance the oscillations became prominent and stable summarized in Table 1.

Table 1. Pumping data for different working fluid volume cases.

Liquid Volumes (mL)	Pumped Volume (mL)	Time (s)
100	20	22
80	8	25

The analyzation of the graphs showed that the temperature variations for the cold end remained similar in both 80 mL and 100 mL working fluid volumes. This is strongly justified and implied by the constant heat input to the hot end in the system. The pressure variations for the 100 mL volume case lye in the range 108–114 kPa while for the 80 mL volume case the variation majorly existed between 102–107 kPa. This implies that although the pressure varied by equal values in the cold end, the magnitude of the pressure trend in the cold end for the 100 mL case was more than that for the 80 mL case. This is also reflected and supported by the pumping data regarding both cases.

Frequencies of oscillations for both cases were similar; however, for the case of 100 mL, the amplitudes were higher with respect to the 80 mL oscillations. The frequency of the oscillations remained a factor of engine geometry and heat input.

For a lower volume of water engine, the starting time was less than that for a higher volume of water; hence, for a lower temperature differential, the amount of working gas should be higher compared to that of working liquid. A similar outcome was proven by Mahdy [10].

3.2. *Effects of Liquid Column Vaporization*

The next part of the research is to notice the effects on engine performance by the vaporization of liquid. For this purpose, an alcohol that is Ethylene Glycol is used to achieve variations of vapor pressure to know the relationship between fluid vaporization and engine performance. Vapor pressure was determined using Raoult's law after the fluids were mixed volumetrically, as illustrated in Tables 2 and 3.

$$P_{solution} = X_{solvent} \cdot P_{solvent} \tag{1}$$

$$X_A = \frac{n_A}{n_A + n_B} \tag{2}$$

Table 2. Heat of vaporization (kJ/mol) and vapor pressure.

Liquid	Heat of Vaporization	Vapor Pressure (100 °C)
Water	40.66	87,726
Ethylene Glycol	65.60	2466

Table 3. Calculated vapor pressure.

Mixtures	Ratio	Vapor Pressure
Glycol, Water	20–80	70,180
Glycol, Water	40–40	43,863
Glycol, Water	50–10	14,621

Temperature and pressure values were recorded for three mixtures (100 mL, 80 mL and 60 mL) at cold end, as illustrated in Figures 8 and 9.

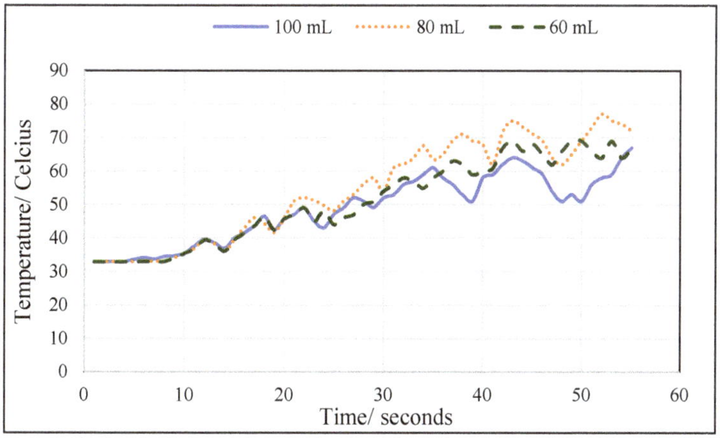

Figure 8. Temperature vs. time Graph for glycol mixtures.

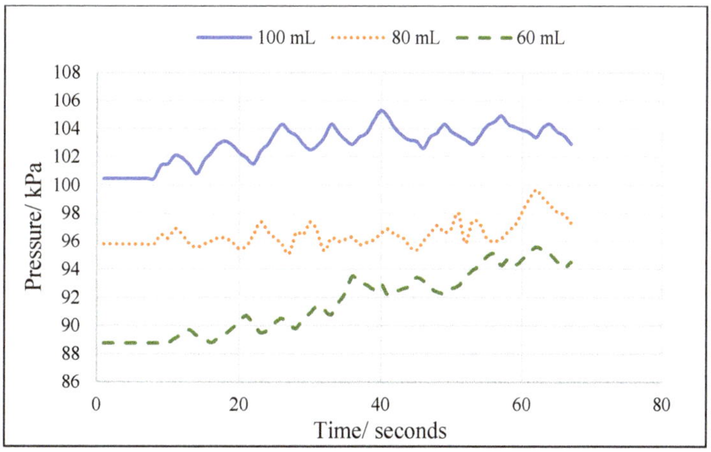

Figure 9. Pressure vs. time Graph for glycol mixtures.

The oscillations in the tuning column were also analyzed using tacker software for all the concentrations for the sake of comparison of oscillations under different operating conditions, as shown in Figures 10–12.

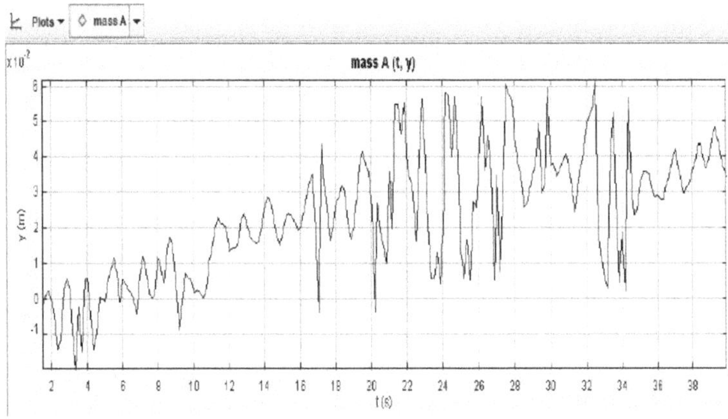

Figure 10. 100 mL mixture.

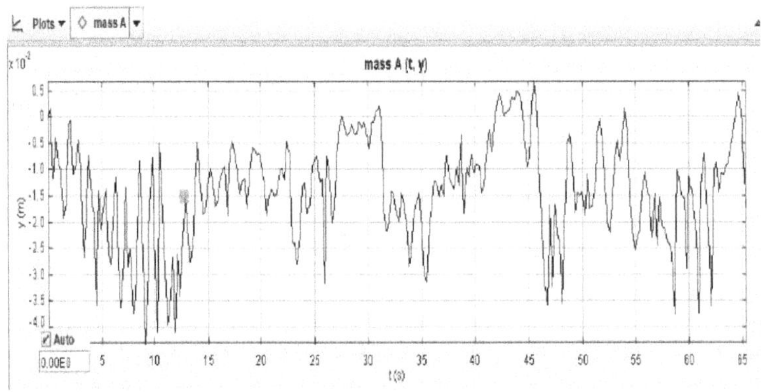

Figure 11. 80 mL mixture.

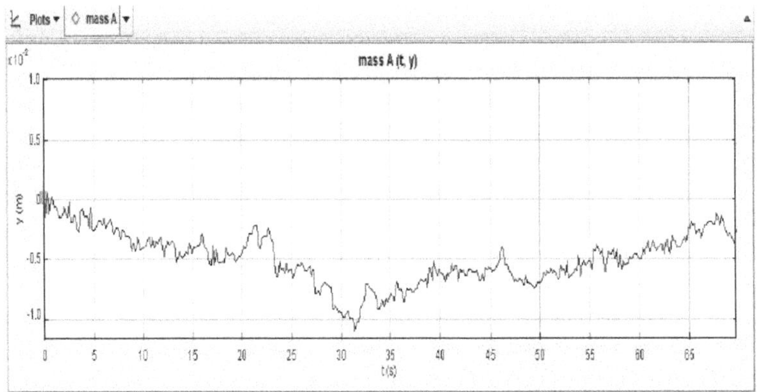

Figure 12. 60 mL mixture.

In all the cases, the source of heat input and conditions were kept the same, and experimental results show that although the range of pressure variation is the same, higher pressures are achieved with greater concentration of water than ethylene glycol. The temperature plots do not show a recognizable or definite trend, possibly because the heat input is maintained constant. The analysis of amplitude vs. time graphs from video tracker modelling and analysis software show that when the liquid columns are occupied with greater concentration of liquid, with higher heat of vaporization, the oscillations are somewhat more stable but with a much smaller amplitude. Compared to the mixtures with greater concentration of water in liquid columns, the graphs show that for a given hot column temperature, the rate of energy conversion into vaporization is higher for liquids having relatively low enthalpies of vaporization, which results in a larger oscillation amplitude. It was found that the engine's operating frequency was reliant on geometry but unaffected by input energy back when the working liquid was just water. The research indicated that there is some dependence of the engine's operating frequency on the composition of glycol mixtures. Similar research was conducted by Newlan, and the results were consistent with previous studies

4. Conclusions

There is a strong correlation between engine output power and a high vapor pressure of the engine working fluid, suggesting that at a constant heat input, as the concentration of the lower heat of vaporization liquids rises, more energy may be transferred to mechanical work. This is in line with the results of Oyewunmi's research.

Author Contributions: Conceptualization, A.R.M. and H.Z.K.; methodology, A.R.M. and M.K.K.; software, A.R.M. and M.H.Y.; validation, A.R.M. and M.H.Y.; formal analysis, A.R.M. and A.A.; investigation, M.K.K.; data curation, A.A.; writing—original draft preparation, A.R.M.; writing—review and editing, M.K.K.; supervision, M.K.K.; project administration, A.R.M. All authors have read and agreed to the published version of the manuscript.

Funding: This research received no external funding.

Acknowledgments: We are thankful to National University of Science and Technology (NUST) for providing work space for conducting experiment.

Conflicts of Interest: The authors declare no conflict of interest.

References

1. Newlan, S.; Sulchek, T.; Telenko, C.; Yee, S. Optimization, and Application of the Liquid Piston Heat Engine through Design and Empirical Modelling. Available online: https://smartech.gatech.edu/bitstream/handle/1853/60132/NEWLAN-THESIS-2017.pdf?sequence=1 (accessed on 23 May 2017).
2. Wang, K.; Sanders, S.R.; Dubey, S.; Choo, F.H.; Duan, F. Stirling Cycle Engines for Recovering Low and Moderate Temperature Heat: A Review. *Renew. Sustain. Energy Rev.* **2016**, *62*, 89–108. [CrossRef]
3. Orda, E.; Mahkamov, K. Development of "Low-Tech" Solar Thermal Water Pumps for Use in Developing Countries. *J. Sol. Energy Eng.* **2004**, *126*, 768–773. [CrossRef]
4. Gupta, A.; Sharma, S.; Narayan, S. *Liquid Piston Engines*; Wiley: Hoboken, NJ, USA, 2017.
5. Cengel, Y.; Boles, M. *Thermodynamics: An Engineering Approach*, 8th ed.; McGaw-Hill Education: New York, NY, USA, 2014.
6. Mason, J.W.; Stevens, J.W. Design and Construction of a Solar-Powered Fluidyne Test Bed. In Proceedings of the ASME 2011 International Mechanical Engineering Congress & Exposition, Denver, CO, USA, 11–17 November 2011.
7. Kyei-Manu, F.; Obodoako, A. Design and Development of a Liquid Piston Stirling Engine. Available online: https://www.swarthmore.edu/sites/default/files/assets/documents/engineering/FK_AO_Final.pdf (accessed on 19 January 2022).
8. Narayan, S.; Singh, R.; Rana, R. Performance Analysis of Liquid Piston Fluidyne Systems. *Mech. Test. Diagn.* **2015**, *5*, 12–18.
9. Oyewunmi, O.A.; Kirmse, C.J.W.; Haslam, A.J.; Müller, E.A.; Markides, C.N. Working-Fluid Selection and Performance Investigation of a Two-Phase Single-Reciprocating-Piston Heat-Conversion Engine. *Appl. Energy* **2017**, *186*, 376–395. [CrossRef]
10. Motamedi, M.; Ahmadi, R.; Jokar, H. A Solar Pressurizable Liquid Piston Stirling Engine: Part 1, Mathematical Modeling, Simulation and Validation. *Energy* **2018**, *155*, 796–814. [CrossRef]

engineering proceedings

Proceeding Paper

To Investigate Thermal Conductivity of Phase Change Material by Incorporating Surfactants †

Qaiser Azim *, Shahid Mehmood and Azhar Hussain

Department of Mechanical Engineering, University of Engineering and Technology Taxila, Punjab 47050, Pakistan
* Correspondence: azimqaiser59@gmail.com

† Presented at the 2nd International Conference on Advances in Mechanical Engineering (ICAME-22), Islamabad, Pakistan, 25 August 2022.

Abstract: Phase change materials (PCMs) store or release heat during their phase transition. Their low thermal conductivity affects the performance of the latent heat energy storage system. This study aimed to investigate the thermal conductivity of PCMs by mixing surfactants. Surfactants with varying weights of sodium stearate (SS), gum arabic (GA), and sodium stearoyl lactylate (SSL) were mixed with paraffin wax. The charging and discharging rate of pure paraffin wax was compared to surfactant-mixed paraffin wax. Pure paraffin wax peaked at 66.76 °C. SSL's highest temperature was 68.3 °C at the critical micelle concentration (CMC) (the concentration of surfactants above which micelles form) and 69 °C above the CMC. GA reached 70 °C at the CMC and 71 °C above the CMC. Sodium stearate (SS) reached 72.12 °C at the CMC and 73.8 °C above the CMC. The melting time test revealed that a reduced melting time resulted in the higher heat conductivity of paraffin. When compared to the melting times of pure paraffin, those of composite PCM containing SSL (>CMC), GA (>CMC), and SS (>CMC) decreased by 6.12%, 14.28%, and 22.44%, respectively. The results show that a concentration above the critical micelle concentration of sodium stearate leads to the best results compared to other samples. Surfactants form micelle, which acted as the conducting medium inside the paraffin wax. Thus, paraffin wax's thermal conductivity was boosted, and the study's goal was met.

Keywords: PCM; paraffin wax; thermal conductivity; surfactants

Citation: Azim, Q.; Mehmood, S.; Hussain, A. To Investigate Thermal Conductivity of Phase Change Material by Incorporating Surfactants. *Eng. Proc.* **2022**, *23*, 1. https://doi.org/10.3390/engproc2022023001

Academic Editors: Mahabat Khan, M. Javed Hyder, Muhammad Irfan and Manzar Masud

Published: 20 September 2022

1. Introduction

Energy is a human need that is decreasing. Energy storage balances supply and demand. It is possible to store energy in a thermal storage system. The ability to store heat reduces the amount of energy required. Phase change materials (PCMs) can store thermal energy at a constant temperature throughout phase shift, making them excellent for solar energy storage, heating, and cooling. However, they have limited thermal conductivity, restricting their practical use in storage systems.

Different strategies are used to improve the phase change material's thermal conductivity. High-thermal-conductivity fillers improve a PCM's thermal conductivity. Metal particles, carbon fibers, extended graphite (EG), and nanoparticles improve PCM thermal conductivity but have downsides.

L. Xia et al. [1] made composite phase change materials with EG. The researchers found that EG networks became more compact as their mass fraction increased. In composite PCMs, increasing the EG content to 10% boosted thermal conductivity tenfold over pure paraffin. EG has a negative effect on a PCM's latent heat, which lowers as the EG fraction increases.

Areeg Shama et al. [2] enhanced paraffin wax's thermal properties by adding CuO nanoparticles in varying amounts. Their experiments showed that 1% paraffin wax/CuO is optimal. Composite PCM slows paraffin melting by 22.22 per cent.

Hadi Fauzi et al. [3] added surfactants to eutectic PCMs. Sodium myristate, sodium palmitate, and sodium stearate at 0, 5, 10, 15, and 20% formed PCM eutectic mixes. Surfactant additions affected the eutectic mixtures' thermal properties and conductivity. The increased latent heat of fusion and thermal conductivity can be achieved by adding 5 percent SM, 5 percent SP, and 5 percent SS to an MA/PA eutectic mixture.

These methods had issues with nanoparticle segregation, leakage, and stability, as well as the poor dispersion of nanoparticles. In this study, we employed surfactants as nanoparticles. Surfactants improve the heat conductivity of paraffin wax because they form micelles at concentrations above the critical micelle concentration.

2. Materials and Methods

Three surfactants, sodium stearate (SS), gum arabic (GA), and sodium stearoyl lactylate (SSL), were chosen as nanoparticles. Six samples were generated by mixing three surfactant particles at and above the critical micelle concentration (CMC) with paraffin wax.

Sample Preparation

We placed 100 g of paraffin wax in a 250 mL glass beaker and heated at 75 °C as shown in Figure 1. Molten paraffin wax was mixed with surfactants at or above the critical micelle concentration. The magnetic hot-plate stirrer was then turned on at 75 °C. Surfactant and paraffin wax were mixed in a beaker by spinning a stirring bar. Magnetic pellets agitated the solution at 1500 rpm for 15 min when one gram of surfactant was added.

Figure 1. Magnetic stirring of samples in the beaker placed on a hot-plate heater.

3. Experimentation Procedure

As shown in Figure 2, the heat sink was filled with composite paraffin wax. One end of a K-type thermocouple was attached to the DAQ controller and the other to the heat sink. The heat sink was positioned at the top of the hot-plate heater in this configuration. The heater for the hot plate was turned on. The signal was transmitted to the data logger using a K-type thermocouple sensor. A data collecting system monitored sample temperatures at multiple positions of the heat sink every minute. During the charging cycle (melting process), paraffin wax was heated for up to 90 min on a hot-plate heater in a testing chamber. The temperature was set at 75 °C. The hot-plate heater was turned off, enabling the composite paraffin wax to cool in the air for up to 90 min (discharging). Charging was the absorption of heat to heat and melt the paraffin wax. Discharging was the release of heat to solidify or cool the material to assess its thermodynamic and thermal heat release rate over time.

(**a**) (**b**)

Figure 2. (**a**) Hot-plate heater melts paraffin wax; (**b**) data logger with a computer system records the temperature of paraffin wax and samples during heating and cooling.

4. Results and Discussion

Three different surfactants with concentrations equal to and greater than the critical micelle concentration (CMC) were mixed with paraffin wax. Two samples of each surfactant-based paraffin wax were prepared at and above the CMC. A data acquisition system recorded the charging and discharging behaviour of these samples. First, the charging and discharging behaviour of only pure paraffin wax was recorded and then compared with surfactant-based paraffin wax samples.

Critical Micellar Concentration Measurement

The CMC of a surfactant was determined by measuring the pH of the solution while adding surfactant particles to molten paraffin wax. We observed that two separate trend lines were found when the pH of the solutions was plotted against the surfactant concentrations. The CMC value relates to the point where the pH rate begins to change. The CMC of each surfactant was calculated using graphs, as shown in the Figure 3. The critical micellar concentrations for SSL, GA, and SS were 10 mM (Molar concentration),12 mM, and 15 mM, respectively.

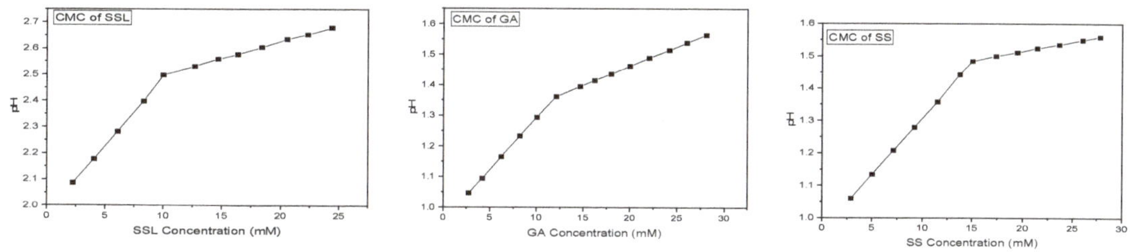

(a) (b) (c)

Figure 3. (**a**) CMC of SSL; (**b**) CMC of GA; (c) CMC of SS.

The charging and discharging results of the samples above the CMC showed a better heat transfer rate than those at the CMC. Figure 4 shows that the maximum peak temperature obtained at the CMC was 68.3 °C and above the CMC was 69 °C in the case of sodium stearoyl lactylate (SSL). In the case of gum arabic (GA), the maximum peak temperature obtained at the CMC was 70 °C and above the CMC was 71 °C. In the case of sodium stearate (SS), the maximum peak temperature obtained at the CMC was 72.12 °C and above the CMC was 73.8 °C. Figure 4 indicates that the PCM's peak temperature increased after the addition of the surfactants. The highest heat transfer occurred in the sodium stearate (SS) mixed with paraffin wax. For the sodium stearate (SS)-based sample, the phase change temperature was 57.7 °C, and the peak temperature was recorded as 72.12 °C at the CMC concentration. A higher peak and phase temperature was recorded at 73.8 °C and 57.13 °C, respectively, for the concentration greater than the CMC.

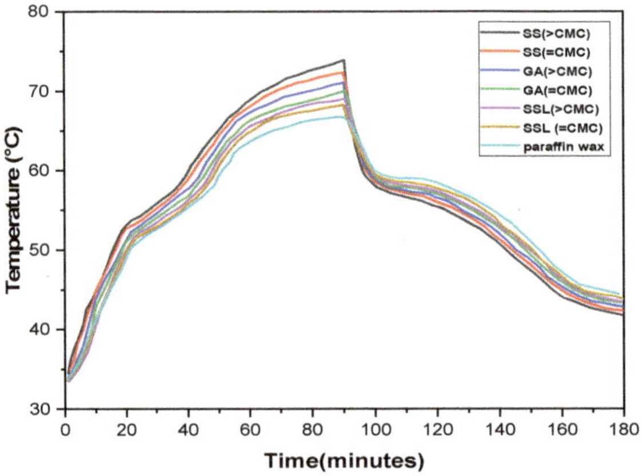

Figure 4. Comparison of charging and discharging profiles of samples with pure paraffin.

Figure 4 also compares the charging and discharging of pure paraffin wax with composite paraffin wax. It shows that composite paraffin wax melts and solidifies faster than pure paraffin wax. The peak temperature of the sodium stearoyl lactylate (SSL) composite increased by 2.30677% and 3.3553% compared to pure paraffin wax at and above the CMC. For the gum arabic (GA) composite, the peak temperature increased by 4.85321% and 6.35111% at and above the CMC. The peak temperature increased by 8.02876% and 10.5452% in sodium stearate (SS) composite at and above the CMC. The increase in paraffin's thermal conductivity was evaluated by contrasting the melting times of composite PCMs with those of paraffin. The melting time was measured from temperature curves until the PCMs reached their melting point (58 °C) from the starting temperature (34 °C). Pure paraffin and the composite PCMs with SSL (=CMC), SSL (>CMC), GA (=CMC), GA (=>CMC), SS (=CMC), and SS (>CMC) melted in 49, 47, 46, 44, 42, 40, and 38 min, respectively, indicating that the composite PCM melting times lowered by 4.08%, 6.12%, 10.20%, 14.28%, 18.36%, and 22.44%, respectively, compared to pure paraffin. The composite PCMs melted faster due to the paraffin's improved heat conductivity. The thermal conductivity of pure paraffin and the composite PCMs with SSL (=CMC), SSL (>CMC), GA (=CMC), GA (>CMC), SS (=CMC), and SS (>CMC) was 0.2 $Wm^{-1}K^{-1}$, 0.22 $Wm^{-1}K^{-1}$, 0.24 $Wm^{-1}K^{-1}$, 0.27 $Wm^{-1}K^{-1}$, 0.3 $Wm^{-1}K^{-1}$, 0.33 $Wm^{-1}K^{-1}$, and 0.36 $Wm^{-1}K^{-1}$, respectively.

5. Conclusions

This study experimented with the charging and discharging of paraffin wax and composite paraffin wax (surfactant). The thermocouples detect the increase in peak temperature during the phase transition process during a specified heating period. Surfactants at different concentrations generate micelles, which act as a heat-transmitting medium in the PCM.

Author Contributions: Conceptualization, Q.A. and A.H.; methodology, Q.A.; software, Q.A.; validation, S.M. and A.H.; formal analysis, Q.A.; investigation, Q.A.; data curation, Q.A. writing—original draft preparation, Q.A.; writing—review and editing, Q.A. and S.M., and A.H.; supervision, S.M.; project administration, Q.A. All authors have read and agreed to the published version of the manuscript.

Funding: This research received no external funding.

Institutional Review Board Statement: Not applicable.

Informed Consent Statement: Informed consent was obtained from all subjects involved in the study.

Data Availability Statement: Not applicable.

Conflicts of Interest: The authors declare no conflict of interest.

References

1. Xia, L.; Zhang, P.; Wang, R.Z. Preparation and Thermal Characterization of Expanded Graphite/Paraffin Composite Phase Change Material. *Carbon* **2010**, *48*, 2538–2548. [CrossRef]
2. Shama, A.; Kabeel, A.E.; Moharram, B.M.; Abosheiasha, H.F. Improvement of the Thermal Properties of Paraffin Wax Using High Conductive Nanomaterial to Appropriate the Solar Thermal Applications. *Appl. Nanosci.* **2021**, *11*, 2033–2042. [CrossRef]
3. Fauzi, H.; Metselaar, H.S.C.; Mahlia, T.M.I.; Silakhori, M.; Nur, H. Phase Change Material: Optimizing the Thermal Properties and Thermal Conductivity of Myristic Acid/Palmitic Acid Eutectic Mixture with Acid-Based Surfactants. *Appl. Therm. Eng.* **2013**, *60*, 261–265. [CrossRef]

engineering proceedings

MDPI

Proceeding Paper

Passive Temperature Excursion of Electronic Devices Using Ionic Wind †

Muhammad Mubashir Iqbal * and Abid Hussain

Department of Mechanical Engineering, Faculty of Mechanical & Aeronautical Engineering, University of Engineering and Technology, Taxila 47050, Pakistan
* Correspondence: muhammadmubashar928@gmail.com

† Presented at the 2nd International Conference on Advances in Mechanical Engineering (ICAME-22), Islamabad, Pakistan, 25 August 2022.

Abstract: Keeping the temperature within safe limits and overcoming the component failures are highly dependent upon heat dissipation in electronic devices. Temperature reduction is a critical factor to be improved in electronic devices due to their compactness in size. Here, a nail-to-cylinder ionic wind generator was constructed for heat reduction. Ionic wind generation represents a new type of cooling for electronic devices. We measured the discharge current and ionic wind speed for various electrode temperatures, electrode distances, and applied voltages. The electrode space and voltage were changed under different conditions; ionic wind velocity was examined within the device. According to the experiments, the installed ionic wind generators represent an effective method of cooling for electronic devices. The experimental results showed that an electrode space of 10.0 mm, a diameter of 25.4 mm, and a voltage of 5 kV led to the maximum ionic wind velocity.

Keywords: passive temperature excursion; nail-to-cylinder design; ionic wind velocity and corona discharge; electronic devices

Citation: Iqbal, M.M.; Hussain, A. Passive Temperature Excursion of Electronic Devices Using Ionic Wind. *Eng. Proc.* **2022**, *23*, 17. https://doi.org/10.3390/engproc2022023017

Academic Editors: Mahabat Khan, M. Javed Hyder, Muhammad Irfan and Manzar Masud

Published: 20 September 2022

1. Introduction

Ionic wind has the potential to develop electro-hydrodynamic thrusters, plasma actuators, and cooling devices. Ionic wind generators can be used to cool small electronic devices in addition to their motionlessness, silence, and compactness. Estimating the mechanical forces produced by the discharge has been a key focus in the development of ionic wind applications. Thermal management and flow control are potential applications of the nail-to-cylinder-type generators [1]. Our high performance could be achieved by combining multiple generators and exploiting the high permeability of the collector electrodes. The application of ionic wind generation cooling has been restricted because of unresolved parameters [2]. There are several parameters to consider, such as the highest possible wind speed, the lowest possible operational voltage, the lowest possible ozone generation, electrode degradation, and the best electrode geometry [3]. Curved electrode tips, semicircular contours to collect electrons [4], and a high-curvature electrode tip for corona discharge have all been proposed in recent studies as ways to reduce voltage requirements [5]. In recent years, more research has been conducted on the use of ionic wind to cool electronic devices.

2. Experimental Setup

Figure 1 depicts an ionic wind generator with a nail-to-cylinder design. Copper nails with 0.7 mm diameters and copper tubes with 1.4 mm diameters were used to make the collection electrode and the electrode for the collectors [6]. The distance between electrodes in this experiment was manually adjusted on a platform where each electrode was attached. A DC power supply was used to measure the voltage and discharge current flowing between the electrodes. Diameter, D, length, L, voltage, V, between electrodes, and distance,

G, between electrodes all have a direct impact on the speed of ionic wind [7]. A range of 5 kV to 10 kV and interelectrode space of −10 mm to 20 mm was used to test cylinder electrodes with diameters 25.4 mm and lengths of 10 mm, 30 mm, and 50 mm [8]. A negative reading meant that the needle electrode had been successfully inserted in series with the cylinder electrode. Measurements were taken in accordance with the procedure outlined below [8]. A 20 s period of operation was required after turning on the power to stabilize the ionic wind flow. Analysis of the average values was conducted for each of the three measurements involving the same control parameters (Table 1).

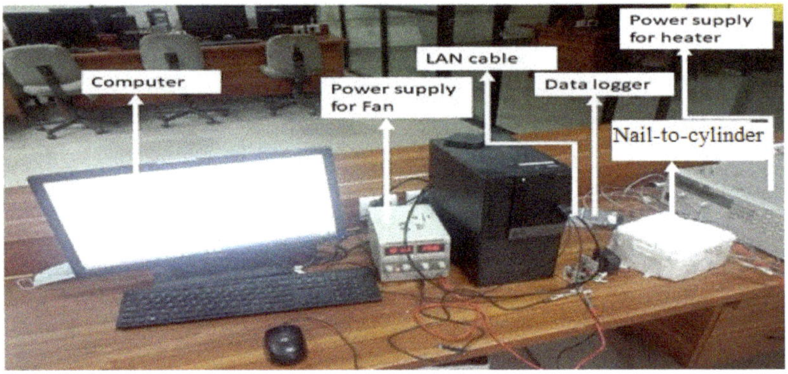

Figure 1. Actual view of the experimental setup.

Table 1. Experiments were conducted with the following input parameters.

Parameters	Test Values	Unit
Voltages applied (V)	5, 6, 7, 8, 9, 10	kV
Electrode distances (G)	−10, −5, 0, 5, 10, 15, 20	mm
Cylinder electrode lengths (L)	10, 20, 30	mm
Cylinder electrode diameters (D)	25.4	mm

3. Results and Discussion

Experimental Results Validation

The motivation to select these parameters was that voltage and discharge current could be changed by manipulating the flow of speed, "U", the geometrical configuration, "G", and the diameter, "D", and by modifying L. As voltages were applied to this device, the ionic wind speed changed from -10 mm to 20 mm with a length value of 30 mm and a diameter of 25.4 mm, as presented in Figure 2a. Figure 2a presents the best results obtained for an available distance (G) between the electrodes, by increasing the applied voltage value and the velocity. By increasing the applied voltage (V), the electric field (E) increased; thus, velocity increased by increasing the voltage applied. Maximum wind velocity was achieved at an electrode distance between 5 mm and 10 mm. The maximum ionic wind velocity was achieved with a ratio of interelectrode distance to cylindrical diameter (G/D ratio) ranges between 0.3 and 0.5 mm.

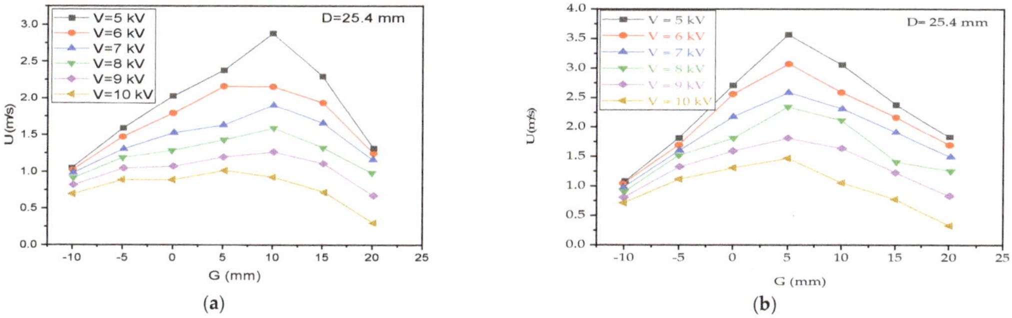

Figure 2. The relationships between applied voltage (V), interelectrode space (G), and ionic wind velocity (U). (**a**) Experimental results (**b**) Literature results.

The output extracted results presented in Figure 2a were validated with the experimental results from Longnan Li [9] displayed in Figure 2b. The presented results show the best validation through comparisons with experimental studies in the literature.

Figure 3 depicts the relationship between current and voltage for an ionic wind generator. To avoid unstable current behavior, we only evaluated electrode lengths that are more than or equal to the distance between the electrodes. Additionally, there was no correlation between the discharge current or other parameters such as the electrode length or ionic wind velocity in the experiments using 10 mm and 50 mm cylindrical electrodes.

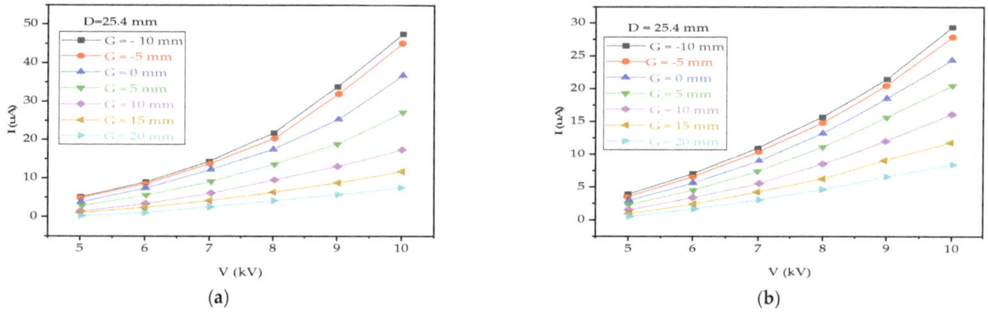

Figure 3. The discharge current–voltage applied characteristics of the ionic wind generator. (**a**) Experimental results (**b**) Literature results.

Figure 3a presents that while increasing the voltage value, the current increased, and the current was maximized when the voltage value was 10 kV. These results were obtained at various distance (G) values between the electrodes and a diameter value of 25.4 mm. For negative distance values between the electrodes, the current dependency on the electric potential was unstable; in this case, it was limited to some extent due to the positive distance (G) value between the electrodes. Using a different cylindrical length electrode, there was no such dependency observed on the ionic wind velocity for other measurements.

To validate the experimental results, the findings were compared with those of Longnan Li, presented in Figure 3b. The obtained results best matched with the results in the literature, and in this way, the obtained results were validated.

4. Conclusions

The purpose of this study was to investigate the cooling effect of ionic wind which was developed by nail-to-cylinder electrodes on electronic devices. According to the experimental results, the ionic wind generator exhibited the following characteristics:

optimum ionic wind voltage, discharge current, and ionic wind velocity. From this study, it is considered that a higher voltage applied enhances the strength of ionic wind. If a larger space between electrodes is used, the ionic wind velocity will be lower for the same applied voltage. In contrast, a larger space between electrodes (G) can extend the operating voltage range and generate higher ionic wind velocity. The maximum ionic wind velocity was achieved with a ratio of interelectrode distance to cylindrical diameter (G/D ratio) ranging between 0.3 and 0.5 mm. Ionic wind operation voltage increased from 5 kV to 10 kV when the interelectrode space was increased from -10 mm to 20 mm. It was possible to achieve the maximum ionic wind velocity when the interelectrode space was 10.0 mm, the diameter was 25.4 mm, and the applied voltage was 5 kV.

Author Contributions: Conceptualization, M.M.I.; methodology, M.M.I.; software, M.M.I.; validation, M.M.I.; formal analysis, M.M.I.; investigation, M.M.I.; data curation, M.M.I.; writing—original draft preparation, M.M.I.; writing—review and editing, M.M.I. and A.H.; supervision, A.H.; project administration, A.H. All authors have read and agreed to the published version of the manuscript.

Funding: This research received no external funding.

Institutional Review Board Statement: Not applicable.

Informed Consent Statement: Not applicable.

Data Availability Statement: Not applicable.

Conflicts of Interest: The authors declare no conflict of interest.

References

1. Moreau, E. Airflow Control by Non-Thermal Plasma Actuators. *J. Phys. D Appl. Phys.* **2007**, *40*, 605–636. [CrossRef]
2. Hsu, C.P.; Jewell-Larsen, N.E.; Sticht, C.; Krichtafovitch, I.A.; Mamishev, A.V. Heat Transfer Enhancement Measurement for Microfabricated Electrostatic Fluid Accelerators. In Proceedings of the 2008 Twenty-fourth Annual IEEE Semiconductor Thermal Measurement and Management Symposium, San Jose, CA, USA, 16–20 March 2008. [CrossRef]
3. Jewell-Larsen, N.E.; Ran, H.; Zhang, Y.; Schwiebert, M.K.; Honer, K.A.; Mamishev, A.V. Electrohydrodynamic (Ehd) Cooled Laptop. In Proceedings of the 2009 25th Annual IEEE Semiconductor Thermal Measurement and Management Symposium, San Jose, CA, USA, 15–19 March 2009; pp. 261–266. [CrossRef]
4. Jewell-Larsen, N.E. Optimization and Miniaturization of Electrostatic Air Pumps for Thermal Management. Ph.D. Thesis, University of Washington, Washington, DC, USA, 2013.
5. Schlitz, D.; Singhal, V. An Electro-Aerodynamic Solid-State Fan and Cooling System. In Proceedings of the 2008 Twenty-Fourth Annual IEEE Semiconductor Thermal Measurement and Management Symposium, San Jose, CA, USA, 16–20 March 2008; pp. 46–49.
6. Liang, W.J.; Lin, T.H. The Characteristics of Ionic Wind and Its Effect on Electrostatic Precipitators. *Aerosol Sci. Technol.* **1994**, *20*, 330–344. [CrossRef]
7. Weinberg, F.; Carleton, F.; Kara, D.; Xavier, A.; Dunn-Rankin, D.; Rickard, M. Inducing Gas Flow and Swirl in Tubes Using Ionic Wind from Corona Discharges. *Exp. Fluids* **2006**, *40*, 231–237. [CrossRef]
8. Kim, C.; Park, D.; Noh, K.C.; Hwang, J. Velocity and Energy Conversion Efficiency Characteristics of Ionic Wind Generator in a Multistage Configuration. *J. Electrostat.* **2010**, *68*, 36–41. [CrossRef]
9. Li, L.; Lee, S.J.; Kim, W.; Kim, D. An Empirical Model for Ionic Wind Generation by a Needle-to-Cylinder Dc Corona Discharge. *J. Electrostat.* **2015**, *73*, 125–130. [CrossRef]

engineering proceedings

MDPI

Proceeding Paper

Design of an Indoor Setup for Experimental Investigation of Thermosiphoning Heat Transfer Using Water and Nanofluid for Application in Compound Parabolic Solar Collectors †

Muhammad Taimoor Jahangir *, Muzaffar Ali *, Ozair Ghufran Bhatti, Muhammad Arbaz, Muhammad Irfan and Muhammad Hassan Haider

Department of Mechanical Engineering, Faculty of Mechanical & Aeronautical Engineering, University of Engineering and Technology Taxila, Punjab 47050, Pakistan
* Correspondence: taimoorjahanger@gmail.com (M.T.J.); muzaffar.ali@uettaxila.edu.pk (M.A.)
† Presented at 2nd International Conference on Advances in Mechanical Engineering (ICAME-22), Islamabad, Pakistan, 25 August 2022.

Abstract: The world is moving towards renewable energy sources because of fossil fuel depletion and its adverse environmental impacts. To study the thermosiphoning process using water and nanofluids at different angles of receiver tubes, an indoor experimental setup was designed. The maximum flow rate achieved at a 35° angle was 6.30 mL/s and the maximum outlet temperature achieved was 82.8 °C at a 45° angle using water. The flow rate achieved using Al_2O_3 nanofluid was 8.20 mL/s. The results show that the time to achieve the thermosiphoning was greatly reduced with an enhanced flow rate of 30.1% using nanofluids as compared with water.

Keywords: thermosiphoning; nanofluids; heat transfer; receiver tube; boundary conditions; Boussinesq approximation

Citation: Jahangir, M.T.; Ali, M.; Bhatti, O.G.; Arbaz, M.; Irfan, M.; Haider, M.H. Design of an Indoor Setup for Experimental Investigation of Thermosiphoning Heat Transfer Using Water and Nanofluid for Application in Compound Parabolic Solar Collectors. *Eng. Proc.* **2022**, *23*, 12. https://doi.org/10.3390/engproc2022023012

Academic Editors: Mahabat Khan, M. Javed Hyder and Manzar Masud

Published: 20 September 2022

1. Introduction

In recent decades, many advancements have been made in the field of solar energy technology. Parabolic collectors are the type of solar collectors used to concentrate sun rays on an absorber tube [1]. The acceptance angle of the compound parabolic collector determines the maximum number of sun rays to be concentrated [2].The concentrated light is used to heat the liquid flowing through it. The circulation of fluid in the tubes takes place using an active system that requires a continuous power source. However, thermosiphoning is a passive system used to circulate the fluid using natural convection [3]. The main aim of this research is to design an indoor setup to investigate the use of thermosiphoning for the purpose of pumpless water flow in a compound parabolic collector. The numerical results use the Boussinesq approximation method for incompressible fluids [4]. Boussinesq approximation suggests that the variation in all of the fluid properties other than density is ignored. The work accomplished in this research is the design and fabrication of an indoor setup for experimental thermosiphoning results. Using the experimental results, the flow rate, volume accumulated, and temperatures were calculated. The use of thermosiphoning in solar water heaters using the flat plate solar collectors has already been implemented and is a mature technology [5]. However, thermosiphoning has not yet been accomplished in compound parabolic solar collectors. There is a considerable research gap in achieving thermosiphoning in the case of compound parabolic collectors. This research paper is a contribution to achieving passive fluid flow in a compound parabolic solar collector through the principle of thermosiphoning. This study also focuses on using Al_2O_3-based nanofluids with enhanced thermal properties as working fluids. This research has not been carried out previously and very little literature reviews are available on it.

2. Methodology

2.1. Simulation Analysis

Complete setup and analysis was carried out using ANSYS 2021 R2. This is the free convection heat transfer phenomena, where the motion of fluids is caused by the buoyancy forces arising from variation in the density of fluid with the temperature. This is why we used the Boussinesq method, which is a density-based solver used to achieve thermosiphoning. The geometry, meshing, and boundary conditions are shown in Figure 1.

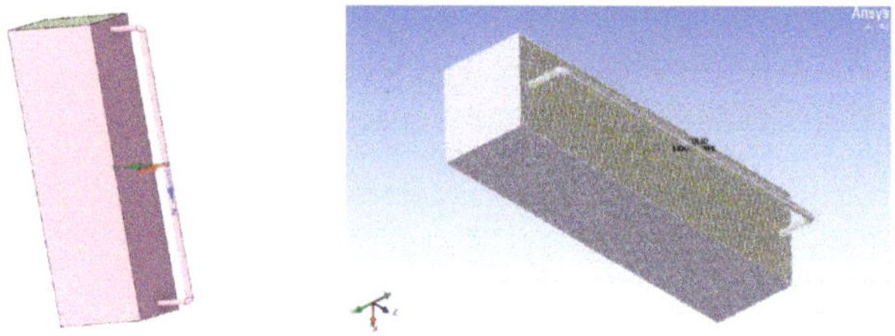

Inlet Velocity	0 m/s
Inlet Pressure	0 Pa
Inlet Temperature	300 K
Heat Flux	300 W/m^2

(a) (b) (c)

Figure 1. (**a**) Ansys model, (**b**) meshing, and (**c**) boundary conditions.

2.2. Indoor Experimentation

The experimentation involved achieving thermosiphoning using water in an indoor designed setup. The heater was made using an induction coil that heats the pipe placed inside of it. The whole setup of the indoor system consists of a heater, copper pipe, water tank, and K-type thermocouple. The schematic diagram and indoor setup diagram are shown in Figure 2.

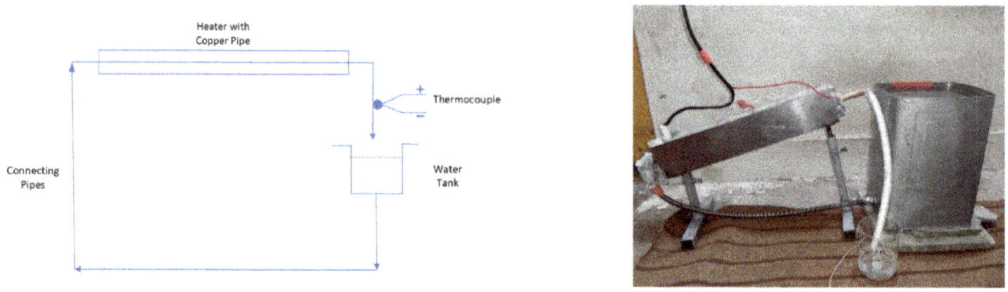

(a) (b)

Figure 2. (**a**) Schematic diagram of the indoor setup and (**b**) indoor experimental setup.

Readings were taken at different angles of inclination. The copper pipe is inside the heating source to gain maximum heat flux through the heater. The heater is made in a way that maximum heat flux reaches the copper tube. The K-type thermocouple is used to measure the temperature at the inlet and outlet of the copper pipe.

3. Results and Discussion

The theoretical results are from the simulation of the pipe on the ANSYS setup 2021 R2. The energy model was used with a viscous laminar type of fluid. The gravity and buoyancy effect was also added into the simulation. The Boussinesq approximation method was used for defining the values of density, specific heat capacity (C_p), thermal conductivity, and viscosity of the fluid with the change in the temperature of the fluid. The simulation results for temperature and velocity are shown in Figure 3.

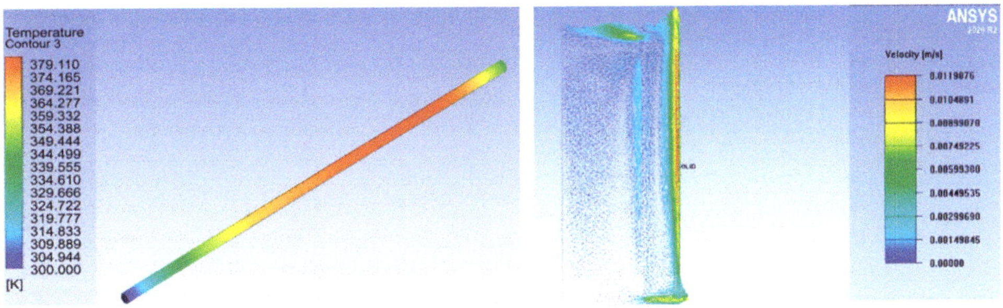

(a) (b)

Figure 3. (**a**) Temperature distribution diagram and (**b**) velocity distribution diagram.

The temperature and velocity distribution diagrams show that the outlet temperature obtained was 379 K, whereas the inlet temperature was 300 K. At the start of the thermosiphoning, the whirling phenomena of fluid occurred, then the continuous flow of fluid was obtained. The maximum outlet velocity obtained was about 0.011 m/s. As these numerical results are close to the experimental results, this shows that, with the increase in the flux value, the outlet temperature and velocity contours also increase. In the experimental setup, three cases of receiver tubes at different angles with water as working fluid are studied. The parameters observed during the experimentation were flow rate, inlet temperature, and outlet temperature. All of the parameters are shown in Table 1 below.

Table 1. Indoor setup specifications and the experimental results.

	Receiver Tube Angles		
	35°	40°	45°
Receiver tube length (in)	21	21	21
Tube diameter (in)	0.63	0.63	0.63
Flow rate achieved (mL/s)	6.30	5.92	4.97
Inlet temperature (°C)	28.5	27	28
Outlet temperature (°C)	76.1	74.9	80.8

The results show that the thermosiphoning starts at different temperatures for all inclination angles. At a 35° angle of the receiver tube, the thermosiphoning starts at a temperature of 54.5 °C, while thermosiphoning start at a slightly higher temperature of 57 °C at an angle of 40°. The longest time and temperature to achieve thermosiphoning took place at 45°, which was 60 °C. This is because a larger temperature difference is required to cause the buoyancy effect. The four cases use a tube diameter of 0.63 in and length of 21 in. The results of the thermosiphoning are shown in Figure 4.

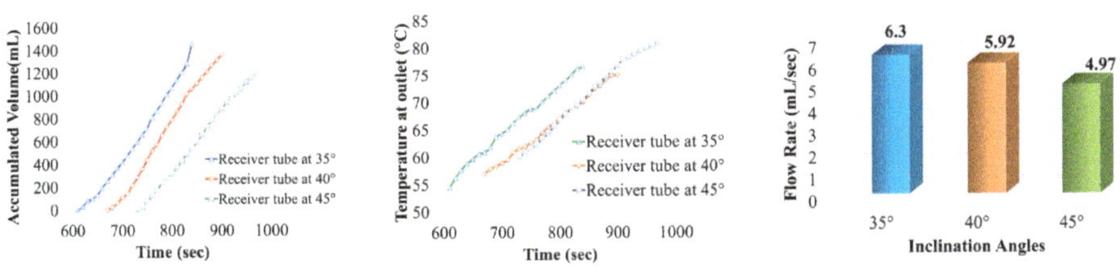

(a) (b) (c)

Figure 4. (**a**) Time vs. accumulated volume, (**b**) time vs. outlet temperature, and (**c**) volume accumulated.

The water-based Al_2O_3 nanofluid with a 0.05% concentration of nanoparticles shows an increased flow rate of 8.2 mL/s at a 35° angle of the receiver tube. The graphs in Figure 5 show that the thermosiphoning took less time and a lower temperature to achieve thermosiphoning as compared with water. The comparison of the numerical and simulation results is also shown in Figure 5 for the case of water as working fluid.

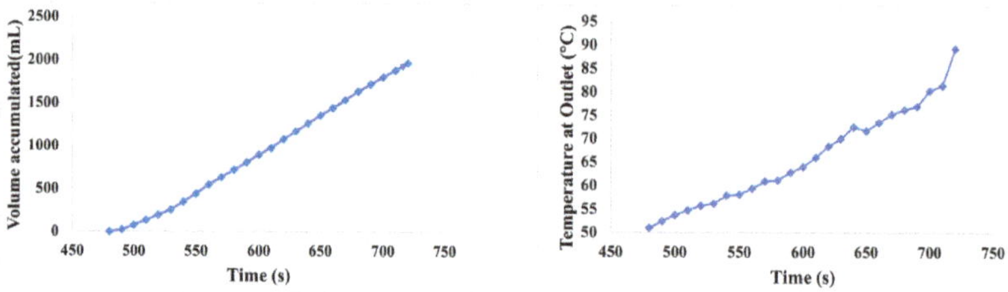

(**a**) (**b**)

Figure 5. (**a**) Volume accumulated with time and (**b**) temperature at outlet.

The maximum flow rate of 6.30 mL/s with water was achieved at a receiver tube angle of 35°. The simulation results show a flow rate of 5.8 mL/s. The simulation shows that an outlet temperature of 379 K was achieved. The temperature at which thermosiphoning begins varied between 10 min and 12 min depending on the angle of the tube. The results obtained using water-based Al_2O_3 nanofluid show that the thermosiphoning time was reduced to just 8 min at a 35° angle. The flow rate achieved was also enhanced by 30% with nanofluids. This difference in time was achieved as a result of the enhanced heat transfer capacity of nanofluids. The results show that water and nanofluids can be employed in compound parabolic collectors for a pumpless water flow if the flow rate requirements are very small.

Author Contributions: Conceptualization, M.T.J. and M.A. (Muzaffar Ali); methodology, M.T.J. and M.I.; software, M.T.J. and O.G.B.; validation, M.T.J. and O.G.B.; formal analysis, M.T.J. and M.H.H.; investigation, M.T.J. and M.I.; data curation, M.A. (Muzaffar Ali); writing—original draft preparation, M.T.J.; writing—review and editing, M.T.J. and M.H.H.; supervision, M.A. (Muzaffar Ali); project administration, M.A. (Muhammad Arbaz). All authors have read and agreed to the published version of the manuscript.

Funding: This research received no external funding.

Institutional Review Board Statement: Not applicable.

Informed Consent Statement: Not applicable.

Data Availability Statement: Not applicable.

Conflicts of Interest: The authors declare no conflict of interest.

References

1. Patel, D.K.; Brahmbhatt, P.K.; Panchal, H. A Review on Compound Parabolic Solar Concentrator for Sustainable Development. *Int. J. Ambient Energy* **2018**, *39*, 533–546. [CrossRef]
2. Chinnasamy, S.; Jothirathinam, S.; Balasubramanian, K.; Anbuselvan, N.; Yuvaraj, T.; Prabaharan, N.; Senjyu, T. Investigation on the Optical Design and Performance of a Single-Axis-Tracking Solar Parabolic Trough Collector with a Secondary Reflector. *Sustainability* **2021**, *13*, 9918. [CrossRef]
3. Ulewicz, M.; Zhelykh, V.; Kozak, K.; Yurii, F. Application of Thermosiphon Solar Collectors for Ventilation of Premises. In Proceedings of the International Conference Current Issues of Civil and Environmental Engineering Lviv-Košice-Rzeszów (CEE 2019), Lviv, Ukraine, 11–13 September 2019; pp. 180–187.

4. Nikitin, L.v.; Ryzhak, E.I. Accuracy of the Boussinesq Approximation for an Incompressible Fluid. *Fluid Dyn.* **1981**, *16*, 174–180. [CrossRef]
5. Ahmad, S.; Ali, M.; Ali, F.; Ahmad, S.; Ahmad, D.; Iftikhar, O. Design and Experimental Investigation of Thermosiphoning Heat Transfer through Nanofluids in Compound Parabolic Collector. *Eng. Proc.* **2021**, *12*, 39. [CrossRef]

engineering proceedings

MDPI

Proceeding Paper

Investigating the Effect of Crack's Inclination on Strain Energy and Stress Intensity under Uniaxial Loading †

Salman Khan [1,*], Massab Junaid [1] and Fahd Nawaz Khan [2]

1 Faculty of Mechanical Engineering (FME), Ghulam Ishaq Khan (GIK) Institute of Engineering Sciences and Technology, Topi 23460, Pakistan
2 Faculty of Material and Chemical Engineering (FMCE), Ghulam Ishaq Khan (GIK) Institute of Engineering Sciences and Technology, Topi 23460, Pakistan
* Correspondence: gme2185@giki.edu.pk
† Presented at the 2nd International Conference on Advances in Mechanical Engineering (ICAME-22), Islamabad, Pakistan, 25 August 2022.

Abstract: The component's strength and reliability are greatly influenced by the occurrence of cracks and flaws. Material surfaces or internal defects can create stress concentration, which causes failure of the materials. To quantify the stress intensity factor (SIF) and strain energy release rate (SERR) around the tips of the crack, numerical simulation has been performed on a plate of central crack with Comsol Multiphysics. The SIF and SERR with varying uniaxial stress loading (5, 10, 15, 20, 25 MPa) and crack's angle (0°–180°) were determined using the J-integral method. The results show that the crack's maximum opening and sliding displacement occurred at an angle of 90° and 45° respectively. For mode I, the SIFs were maximum at the crack's inclination of 90° and for mode II, it was maximum at an angle of 45°. Similarly, the SERR was also found to be maximum when the crack was normal to the applied stress and observed to increase with the increase in applied stress.

Keywords: stress intensity factor (SIF); strain energy density (G); J-integral method; crack's angle; stress concentration; opening; and slide mode displacement

check for updates

Citation: Khan, S.; Junaid, M.; Khan, F.N. Investigating the Effect of Crack's Inclination on Strain Energy and Stress Intensity under Uniaxial Loading. *Eng. Proc.* **2022**, *23*, 9. https://doi.org/10.3390/engproc2022023009

Academic Editors: Mahabat Khan, M. Javed Hyder, Muhammad Irfan and Manzar Masud

Published: 20 September 2022

1. Introduction

There are certain flaws or cracks either interior or at the surface of the components which were created due to high stresses induced during the manufacturing process. These cracks or flaws cause stress intensity when the components are under loading conditions. The stress distribution or intensity needs to be quantified around the tips of the crack because it creates a state of stress concentration (SC). SIF is a fracture mechanic parameter and is used to define the crack tips stress singularity and is denoted by K. When the SIF becomes equal to or greater than the fracture toughness, the component gets fractured and fails. Based on the crack surface displacement, the stresses near the crack tip can be divided into three basic types as shown in Figure 1 [1].

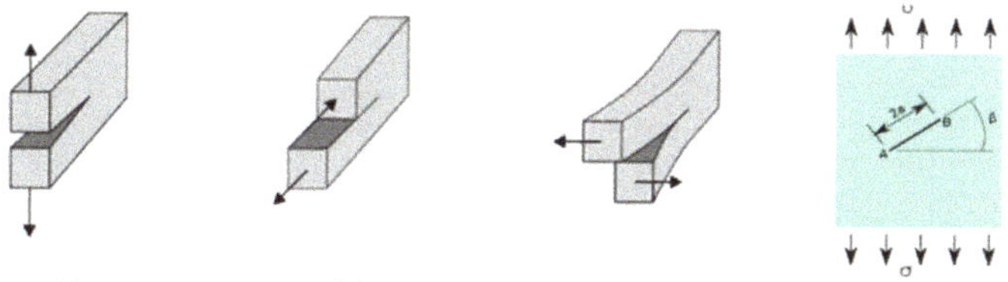

(a) (b) (c) (d)

Figure 1. The three types of loading with respective crack' modes (**a**) Mode I (**b**) Mode II (**c**) Mode III [1] (**d**) The inclination of the crack with the load stress [2].

Research has been performed to develop fracture mechanics criteria to quantify the fracture toughness of different materials and developed different criteria. These fracture criteria can be classified into the following three groups which are further subdivided: stress, strain, and energy-based criterion [3].

For the uniaxial stress loading, the analytical equations for finding the SIFs within a rectangular plate of infinite length can be calculated by [4];

$$K_I = \sigma \times \sqrt{(\pi a)} \times sin^2\beta \quad (1)$$

$$K_{II} = \sigma \times \sqrt{(\pi a)} \times sin\beta \times cos\beta \quad (2)$$

The stress and strain around the crack tip are greatly affected by a constant term called T-stress and can lead to better results [5]. Energy-based theory cannot give a good prediction about the kink angle while the strain energy released rate (SERR) and maximum tangential stress (MTS) criteria give better predictions about the crack growth trajectories [6]. For a semi-infinite plate, the relationships between SIFs and the oblique edge fracture were investigated [7]. An experimental investigation was performed on the plate with a crack edge and find the effect on the SIFs (K_I and K_{II}) [8].

In the paper, a rectangular solid plate with an interior fracture is subjected to uniaxial stress. The fracture or crack is at an angle with the load direction, implying that it exhibits in both modes I and II. The J-integral approach is used to calculate the SERR at crack tips. As the stress intensity was caused at the tip of the crack and these SIFs must be quantified for both mode I and mode II using J- the integral method. The SIFs and SERR, the opening and sliding displacement with the crack's inclination and applied stress was studied using COMSOL Multiphysics software.

2. Numerical Modeling

A rectangular Aluminum plate with a central slanted crack is subjected to a stress in the Y direction. The crack's inclination varies from 0°–180° while the applied stress field varies from 5–25 MPa. The dimension of the plate is given in Table 1. The SIF and the SERR are to be calculated at each angle of the crack. The tips of the crack under the mixed mode (Tensile, compressive and shear) of stress loading can be in opening mode with the tensile mode (denoted by K_I), in plan shear mode (denoted by K_{II}) [9]. For this purpose, the Comsol Multiphysics was used to quantify the effect of loading stress and crack's inclination on the SIFs, SERR, opening and sliding displacement.

Table 1. The dimensions of the rectangular plate with central crack.

S. No	Parameters	Symbol	Values
1	Plate's length	L	100 mm
2	Plate's width	W	80 mm
3	Crack length	2a	40 mm
4	Crack angle	β	0°–180°
5	Load in Y-axis	σ	5–25 MPa

3. Results and Discussion

A 2D rectangular plate with a central crack was studied using Comsol Multiphysics. The load was applied in the uniaxial direction in Y-direction. The crack' angle was varied from 0° to 180° and the respective opening and sliding displacements, the SERR and the corresponding SIFs were determined. The slanted crack can create two types of SIFs, one for mode I (opening) and other for mode II (shear). As the crack is at center of the plate and it has two sharp edges, so at each edges the SIF was calculated for each mode. The variation of the opening and sliding displacement of the crack with the inclination and varying load is shown in Figure 2a,b respectively. Since the plate is two-dimensional plate, there will no tearing mode of crack. The SERR at both the crack tip 1 and tip 2 are given in

Figure 2c,d. The SERR was high when the crack is normal to the applied load. For mode I, the SIFs at both the crack's tip are given in Figure 2e,f. It will be high when the applied load and crack are normal to each other. Similarly, for mode II, the SIF is maximum when the applied load is at 45° to the crack as given in Figure 2g,h. The numerical simulation results are shown in Figure 3.

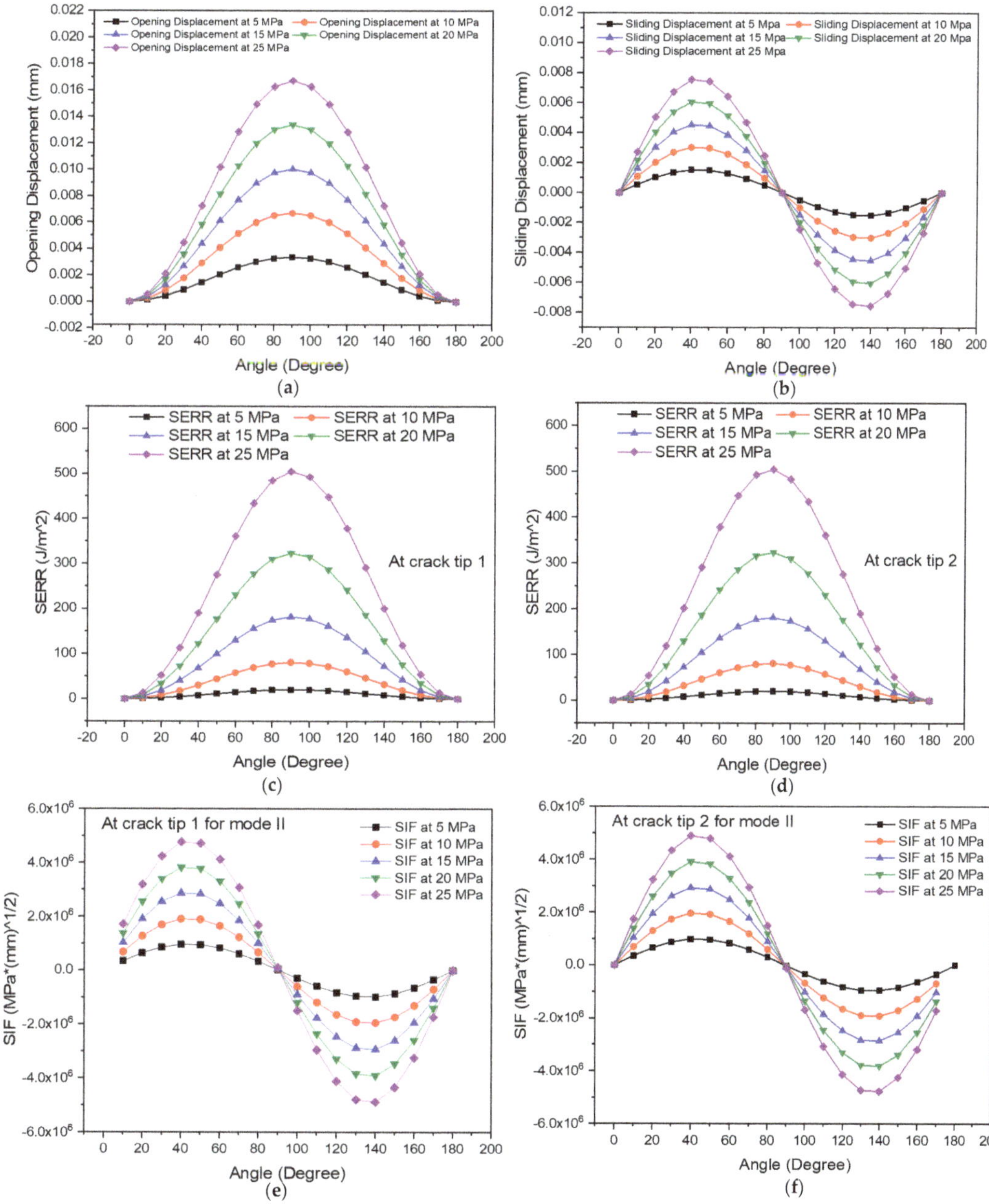

Figure 2. *Cont.*

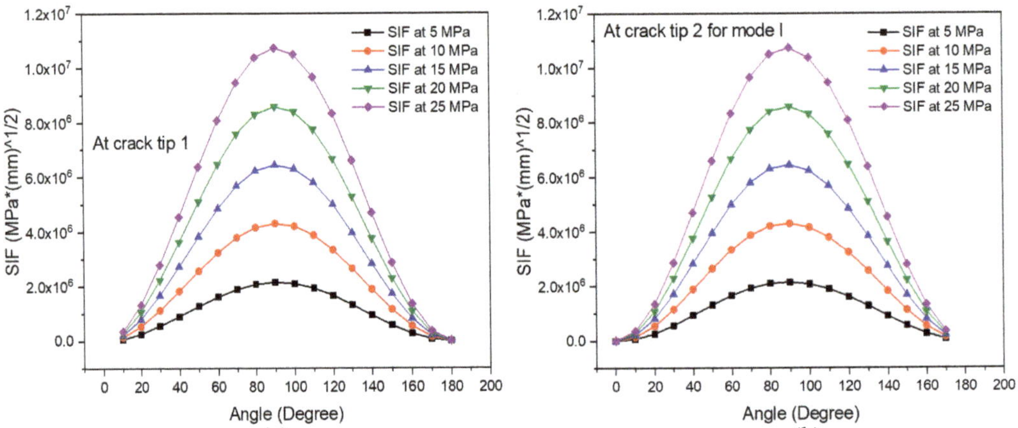

(g) (h)

Figure 2. The variation of crack opening displacement with angle and loads (**a**) The variation of crack sliding displacement with inclination and loads (**b**) The SERR for mode I at crack tip 1 (**c**) The SERR for mode I at crack tip 2 (**d**) The SIF for mode II at crack tip 1 (**e**) The SIF for mode II at crack tip 2 (**f**) The SIF for mode I at crack tip 1 (**g**) The SIF for mode I at crack tip 1 (**h**).

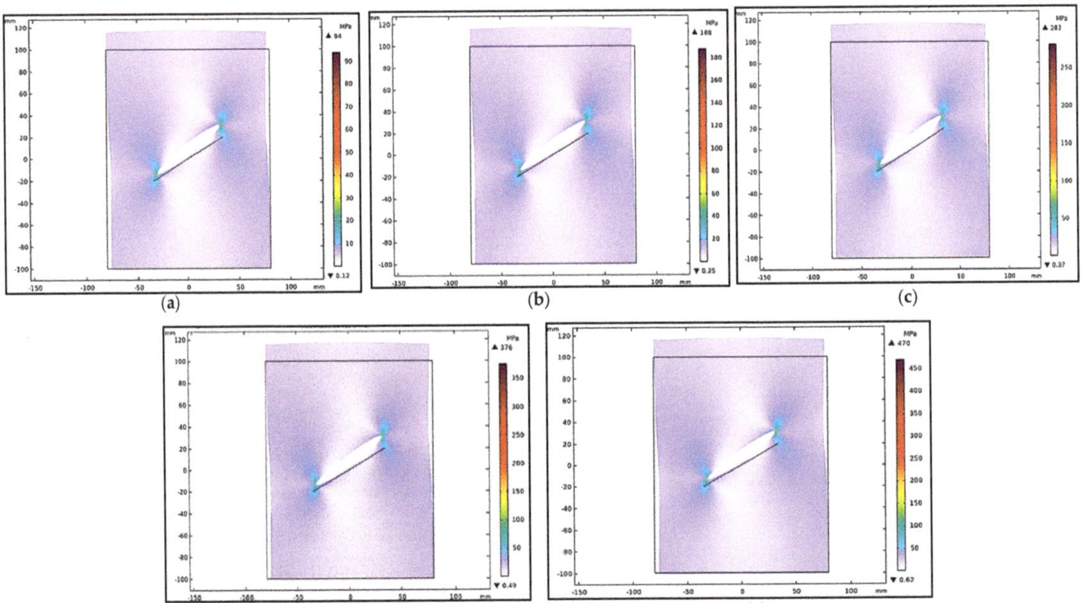

(a) (b) (c) (d) (e)

Figure 3. Von mises stress distribution at different applied loads (**a**) 5 MPa (**b**) 10 MPa (**c**) 15 MPa (**d**) 20 MPa (**e**) 25 MPa.

4. Conclusions

From the numerical investigation, it was concluded that:

4.1. For Mode I (Opening)

i. The opening displacement of the crack was maximum when the stress was applied normally (at 90°) to the crack and increased with the increase in the applied stress field.
ii. The SERR was high when the crack was normal to the applied stress.
iii. SIF was also high when the angle is 90° between applied stress and crack and was directly related to the applied stress.

4.2. For Mode II (Sliding)

i. The crack sliding displacement was maximum when the applied stress and the crack were at an angle of 45° and was directly related to the stress field.
ii. The SERR was high when the crack was at 45° to the applied stress.
iii. SIF was also high when the angle was 45° between applied stress and crack.

Author Contributions: Conceptualization, S.K. and M.J.; methodology, S.K. and M.J.; software, S.K. and F.N.K.; validation, S.K. and F.N.K.; formal analysis, S.K. and M.J.; investigation, S.K. and M.J.; data curation, S.K.; writing—original draft preparation, S.K.; writing—review and editing, S.K., M.J. and F.N.K.; supervision, F.N.K.; project administration, F.N.K. All authors have read and agreed to the published version of the manuscript.

Funding: This research received no external funding.

Institutional Review Board Statement: Not applicable.

Informed Consent Statement: Not applicable.

Data Availability Statement: The data will be available on request.

Conflicts of Interest: The authors declare no conflict of interest.

References

1. Turis, M.; Ivánková, O. Using Finite Element Analysis to Obtain Plastic Zones in the Vicinity of the Crack Edges, under Mixed Mode Loading Conditions. In Proceedings of the 4th International Scientific Conference Structural and Physical Aspects of Construction Engineering (SPACE 2019), Strbske Pleso, Slovakia, 13–15 November 2019; Volume 310. [CrossRef]
2. Zhu, N.; Oterkus, E. Calculation of Stress Intensity Factor Using Displacement Extrapolation Method in Peridynamic Framework. *J. Mech.* **2020**, *36*, 235–243. [CrossRef]
3. Hou, C.; Jin, X.; Fan, X.; Xu, R.; Wang, Z. A Generalized Maximum Energy Release Rate Criterion for Mixed Mode Fracture Analysis of Brittle and Quasi-brittle Materials. *Theor. Appl. Fract. Mech.* **2019**, *100*, 78–85. [CrossRef]
4. Arunkumar, S.; Nithin, V.K. Estimation of Stress Intensity Factor of Multiple Inclined Centre Cracks under Biaxial Loading. *J Fail. Anal. and Preven.* **2020**, *20*, 2040–2058. [CrossRef]
5. Sedighiani, K.; Mosayebnejad, J.; Ehsasi, H.; Sahraei, H.R. The Effect of T-stress on the Brittle Fracture under Mixed Mode Loading. *Procedia Eng.* **2011**, *10*, 774–779. [CrossRef]
6. Sakha, M.; Nejati, M.; Aminzadeh, A.; Ghouli, S.; Saar, M.O.; Dresner, T. On the Validation of Mixed-mode I/II Crack Growth Theories for Anisotropic rocks. *Int. J. Solids Struct.* **2022**, *241*, 111484. [CrossRef]
7. Kim, H.K.; Lee, S.B. Stress Intensity Factors of an Oblique Edge Crack Subjected to Normal and Shear Tractions. *Theor. Appl. Fract. Mech.* **1996**, *25*, 147–154. [CrossRef]
8. Mohsin, N.R. Influence of the Location and Crack Angle on the Magnitude of Stress Intensity Factors Mode i and ii under Uniaxial Tension Stresses. *Eur. J. Mater. Sci.* **2016**, *3*, 64–77.
9. Fajdiga, G. Determining a kink angle of a crack in mixed mode fracture using maximum energy release rate, SED and MTS criteria. *J. Multidiscip. Eng. Sci. Technol.* **2015**, *2*, 356–362.

engineering proceedings

Proceeding Paper

Evaluating the Effect of Process Parameters on the Mechanical Properties of an AA7075-Cu Overcast Joint Using the Taguchi Method †

Muhammad Waqas Hanif *, Ahmad Wasim and Muhammad Sajid

Department of Industrial Engineering, University of Engineering and Technology Taxila, Punjab 47050, Pakistan
* Correspondence: waqas.hanif@students.uettaxila.edu.pk

† Presented at the 2nd International Conference on Advances in Mechanical Engineering (ICAME-22), Islamabad, Pakistan, 25 August 2022.

Abstract: The present study aims to evaluate the effect of process parameters such as melt temperature (MT), squeeze pressure (SP) and insert pre-heating temperature (IPT) on the ultimate tensile strength (UTS) of an AA7075-Cu overcast joint using an orthogonal array (L8) of Taguchi techniques. The analysis of variance (ANOVA) test results showed that SP is the most significant process parameter due to the significant reduction in air gaps between aluminum and copper. The optimal value of UTS (39.33 MPa) was attained at higher levels of SP (90 MPa), MT (800 °C) and IPT (300 °C). The confirmation test validated the significance of process parameters in optimizing the UTS of an AA7075-Cu overcast joint due to the low percentage error.

Keywords: ultimate tensile strength (UTS); orthogonal array (L8); AA7075-Cu overcast joint; insert pre-heating temperature (IPT)

Citation: Hanif, M.W.; Wasim, A.; Sajid, M. Evaluating the Effect of Process Parameters on the Mechanical Properties of an AA7075-Cu Overcast Joint Using the Taguchi Method. *Eng. Proc.* **2022**, *23*, 3. https://doi.org/10.3390/engproc2022023003

Academic Editors: Mahabat Khan, M. Javed Hyder, Muhammad Irfan and Manzar Masud

Published: 20 September 2022

1. Introduction

The application of the (7xxx) aluminum alloy series is quite significant in the automobile and aerospace industries due to its high strength and low density [1]. The properties of these alloys are determined by the selection of casting techniques and the process parameters involved in their manufacturing processes [2]. In this regard, researchers have developed different metal joining techniques and have classified them into three groups: solid with solid bonding, liquid with liquid bonding and solid with liquid bonding. It is worth mentioning that liquid with liquid bonding is economically unfeasible, and solid with solid bonding is a time-consuming process. Hence, these two methods are not feasible for industrial practices.

However, solid with liquid bonding, also called overcasting, has superior characteristics in multisystem applications, such as design flexibility and a low processing cost. Many researchers have studied the impact of control process variables on the mechanical characteristics of overcast joints, such as Ali et al. [3], who observed that, at the die temperature, SP and MT have significant effects on the UTS, and micro-hardness of Al-Al overcast joints. In another article, Ali et al. [4] found that IPT also has prominent effects—followed by SP and pouring temperature—on the UTS and micro-hardness of aged Al-Al overcast joints. Additionally, Liu et al. [5] observed that copper and Al alloy have a high diffusion affinity with each other at high temperatures. However, the effects of IPT, SP and MT on the UTS of the AA7075-Cu overcast joint have not been explored yet. Therefore, the purpose of this research work is to evaluate the effect of process parameters such as SP, IPT and MP on the UTS of the AA7075-Cu overcast joint using the Taguchi technique.

2. Materials and Methods

The chemical composition of the AA7075 material has been confirmed using a spark optical emission spectroscopy test, as shown in Table 1. The copper was found to be 90 percent pure copper. After that, copper inserts were machined to a size of (4 × 13 × 90) mm and polished using abrasive paper to remove any burr. A zinc coating has been performed using electroplating techniques to achieve a 5 µm thickness on a copper insert. It is worth noting that the optimal zinc coating thickness on a solid copper insert was 5 µm [5]. Overcast billets of AA7075-Cu have been manufactured in four stages. In the first stage, AA7075 melts in the electrical resistance furnace to achieve the desired melt temperature. In the second stage, the insert was preheated to achieve the insert pre-heating temperature and inserted into the die block's ejection pin. At the third stage, an oxyacetylene torch was used to achieve a die temperature of 250 °C, and AA7075 was poured into the die block. In the last stage, one operator unclamps the die block, and the other operator disconnects the cast billet from the ejection bin using a hammer. Furthermore, overcast billets are machined at the center of cylindrical billets into rectangular shapes according to the ASTM E8/E8M-11 standard [6].

Table 1. Composition of selected material.

Elements	Mg	Mn	Si	Fe	Cu	Ti	Zn	Cr	Al
Weight Percent (Al)	2.1	0.3	0.4	0.5	1.2	0.2	5.1	0.18	Remainder

3. Experiment Design

Based on the literature review [3,4], and after performing the trial runs, higher and lower levels of process parameters such as SP (60 to 90 MPa), MT (750 to 800 °C) and IPT (250 to 300 °C) were determined. The Taguchi orthogonal L8 array was selected to design the experiment due to three process parameters, and each process parameter has two levels (2^3). Therefore, eight experiments were conducted using Minitab 19.0 software to ensure that all levels of the selected process parameters were considered equally. Additionally, three E8/E8M-11 ASTM standard specimens were extracted from each overcast billet, and three readings of UTS were taken at each experimental setting for repeated measurements and more reliable results. A universal tensile tester with a capacity of 50 KN was used to measure the UTS value at a strain rate of 0.005 mm/s and at an ambient temperature. The mean value of these three measurements was considered as the final value, as shown in Table 2.

Table 2. Experimentation according to the Taguchi orthogonal L8 array.

Experiment No.	Input Variables			Output Response (UTS)	
	SP	MT	IPT	Mean	Standard Deviation
1	60	750	250	16.00	1.50
2	60	750	300	19.00	1.32
3	60	800	250	22.67	1.26
4	60	800	300	27.50	1.32
5	90	750	250	23.63	1.52
6	90	750	300	28.13	1.10
7	90	800	250	32.20	1.18
8	90	800	300	39.33	0.76

4. Results and Discussion

The experiment results showed that an optimal UTS value was achieved at 90 MPa SP, 800 °C MT and 300 °C IPT, and the lowest UTS value was attained at 60 MPa SP, 750 °C

MT and 250 °C IPT, as shown in Table 2. The stress–strain curve at optimal experimental settings is given in Figure 1a. The ranking criteria in the mean ratio depicted that SP was ranked 1 due to the highest contribution ratio (0.48), as shown in Table 3. This was also evident from the scanning electron microscope (SEM) flat specimen (Figure 1b,c). Furthermore, Minitab 19.0 software was used to run an analysis of variance (ANOVA) test to determine the significance of each process parameter. In this analysis, the values of adjusted R^2 and predicted R^2 were 0.997 and 0.998, respectively. These two parameters show that there is less variability between the adjusted and predicted data. Additionally, the standard deviation (SD) was obtained as 0.135, and this indicates the accuracy of the developed model. ANOVA Table 4 shows that SP, MT and IPT have significant effects on the UTS, because the p values are less than 0.05. The main effects plot from Figure 2a shows that increasing SP from 60 to 90 MPa, MT from 750 to 800 °C and IPT from 250 to 300 °C increases the UTS significantly. The low level of SP causes large air gaps between the solid Cu insert and AA7075 that lead to a lower UTS value, and a higher UTS is achieved at a higher level of SP (90 MPa) due to the reduction in air gaps between the solid Cu inserts and AA7075, as shown in Figure 1c. It is also evident from the previous studies that the UTS of the Al-Cu overcast joint increases with increasing MT and SP, and aging parameters improved the UTS of the Al-Al overcast joint by 12.4% [4].

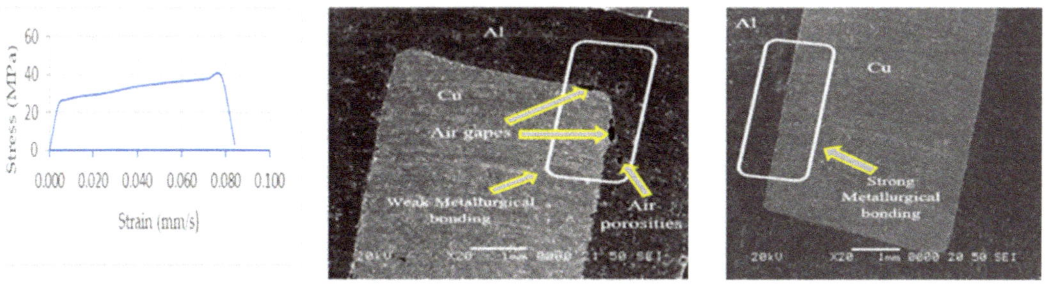

(a) (b) (c)

Figure 1. (**a**) Stress–strain curve at optimal experimental settings; SEM images at (**b**) 60 MPa SP, 750 °C MT and 250 °C IPT; (**c**) 90 MPa SP, 800 °C MT and 300 °C IPT.

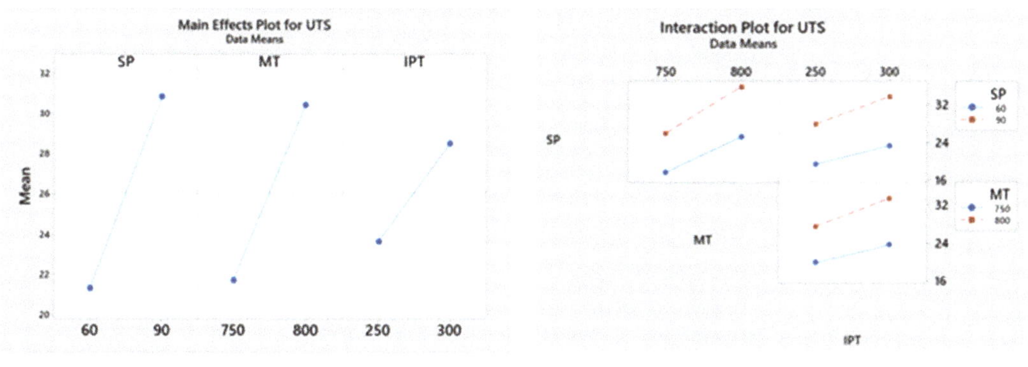

(a) (b)

Figure 2. (**a**) Main process parameter effects plot for mean ratios; (**b**) interactions plot for UTS.

Table 3. Response table for means of selected parameters.

Level	SP	MT	IPT
1	21.29	21.69	23.63
2	30.83	30.43	28.49
Delta	9.53	8.73	4.87
Rank	1	2	3

Table 4. ANOVA for means.

Source	DF.	Seq SS.	Adj SS.	Adj MS.	F	P	Contribution Ratios
SP	1	181.769	181.769	181.769	103.51	0.001	0.48
MT	1	152.542	152.542	152.542	86.87	0.001	0.40
IPT	1	47.369	47.369	47.369	26.98	0.007	0.12
Residual Error	4	7.024	7.024	1.756			
Total	7	388.704					

The interaction plot from Figure 2b depicted that MT and IPT interactions have the most significant effect, since low levels of MT and IPT cause the inappropriate melting of zinc coating, which leads to a reduction in the diffusion coefficient, and a small number of Cu atoms diffuse into the AA7075, as evident from Figure 1b. As a result, a lower UTS was observed at low levels of MT and IPT. Likewise, a higher UTS was attained at higher levels of MT and IPT due to the appropriate melting of zinc coating, which leads to a large number of Cu atoms diffused into the AA7075, as evident in Figure 1c. Consequently, a higher UTS was observed at higher levels of MT and IPT. It is also noted that MT and IPT interactions have significant effects on the squeeze overact joint due to the appropriate melting of zinc coating at higher levels of MT and IPT [4].

5. Validations

To validate the results obtained from the Taguchi method, a confirmatory test has been conducted, as shown in Table 5.

$$X_{predicted} = X_{avg} + (X_{IPT} - X_{avg}) + (X_{SOP} - X_{avg}) + (X_{MOPT} - X_{avg}) \quad (1)$$

In Equation (1), $X_{IPT,}$ X_{SOPT} and X_{MOPT} are the maximum values of IPT (28.49), SP (30.83) and MT (30.43) from Table 3. X_{avg} represents the average UTS value (26.06) seen in Table 2. The actual value of UTS was considered at optimal process parameters. The confirmation test demonstrated the importance of selecting the right combinations of process parameters.

Table 5. Confirmation test.

Actual Value	Predicted Value	Error	Percentage Error
39.33	38.17	1.16	3.63

6. Conclusions

It was concluded that 90 MPa SP, 800 °C MT, 300 °C IPT and MT and IPT interaction were the optimal process parameters to achieve the UTS of an AA7075-Cu overcast joint. Furthermore, it has been observed that an increase in MT, IPT and SP from low levels to high levels significantly increases the UTS value due to an appropriate melting of zinc coating, a large number of copper items diffused into the AA7075 and a reduction in gas porosities that leads to strong metallurgical bonding.

Author Contributions: Conceptualization, M.W.H. and A.W.; methodology, M.W.H. and M.S.; software, M.W.H. and M.S.; validation, M.W.H. and M.S.; formal analysis, M.W.H. and A.W.; investigation, M.W.H. and M.S.; data curation, A.W.; writing—original draft preparation, M.W.H.; writing—review and editing, M.W.H. and M.S.; supervision, M.S.; project administration, M.W.H. All authors have read and agreed to the published version of the manuscript.

Funding: This research received no external funding.

Institutional Review Board Statement: Not applicable.

Informed Consent Statement: Not applicable.

Data Availability Statement: Not applicable.

Conflicts of Interest: The authors declare no conflict of interest.

References

1. Zhou, B.; Liu, B.; Zhang, S. The Advancement of 7xxx Series Aluminum Alloys for Aircraft Structures: A Review. *Metals* **2021**, *11*, 718. [CrossRef]
2. Sajid, M.; Wasim, A.; Hussain, S.; Jahanzaib, M. Manufacturing feature-based cost estimation of cast parts. *China Foundry* **2018**, *15*, 464–469. [CrossRef]
3. Ali, M.; Jahanzaib, M.; Wasim, A.; Hussain, S.; Anjum, N.A. Evaluating the Effects of As-Casted and Aged Overcasting of Al-Al Joints. *Int. J. Adv. Manuf. Technol.* **2018**, *96*, 1377–1392. [CrossRef]
4. Ali, M.A.; Ishfaq, K.; Raza, M.H.; Farooq, M.U.; Mufti, N.A.; Pruncu, C.I. Mechanical Characterization of Aged AA2026-AA2026 Overcast Joints Fabricated by Squeeze Casting. *Int. J. Adv. Manuf. Technol.* **2020**, *107*, 3277–3297. [CrossRef]
5. Liu, T.; Wang, Q.; Sui, Y.; Wang, Q.; Ding, W. An Investigation into Interface Formation and Mechanical Properties of Aluminum–Copper Bimetal by Squeeze Casting. *Mater. Des.* **2016**, *89*, 1137–1146. [CrossRef]
6. Astm, C. *Philadelphia: American Society for Testing Materials*; ASTM Standards: West Conshohocken, PA, USA, 1958.

engineering proceedings

MDPI

Proceeding Paper

A Joint Optimization of Maintenance and Scheduling for Unrelated Parallel Machine Problem Based on Hybrid Discrete Spider Monkey Optimization Algorithm †

Ke Ke and Yarong Chen *

School of Mechanical and Electrical Engineering, Wenzhou University, Wenzhou 325035, China
* Correspondence: yarongchen@126.com

† Presented at the 2nd International Conference on Advances in Mechanical Engineering (ICAME-22), Islamabad, Pakistan, 25 August 2022.

Abstract: In general, the parallel machine scheduling problem that minimizes maximum completion time is NP-hard in a strong sense; a lot of heuristics have been proposed for this kind of problem. In this paper, the unrelated parallel machine scheduling problem with maintainability (UPMSPM) is studied, in which the reliability of machines obeys exponential distribution. A hybrid algorithm HDSMO, which combines the discrete spider monkey algorithm (SMO) with the crossover and mutation operation, is proposed to solve UPMSPM. In view of the lack of local search capability in the later iteration of the traditional SMO algorithm, inertial weights are introduced to update the local leader and the global leader. Computational experiments with randomly generated instances demonstrate that the proposed HDSMO algorithm can obtain significantly better solutions in a shorter time than GA and SMO algorithms.

Keywords: unrelated parallel machine scheduling; spider monkey optimization; preventive maintenance

check for updates

Citation: Ke, K.; Chen, Y. A Joint Optimization of Maintenance and Scheduling for Unrelated Parallel Machine Problem Based on Hybrid Discrete Spider Monkey Optimization Algorithm. *Eng. Proc.* **2022**, *23*, 16. https://doi.org/10.3390/engproc2022023016

Academic Editors: Mahabat Khan, M. Javed Hyder, Muhammad Irfan and Manzar Masud

Published: 20 September 2022

1. Introduction

UPMSP is an important branch of production scheduling. In the real-world production system, long-term running wear and performance degradation of the machines can easily lead to production interruptions, requiring preventive maintenance (PM) to keep machines running [1]. Therefore, it is of great significance to consider the joint optimization of maintenance and scheduling for UPMSP [2]. UPMSP studies considering maintenance are relatively rare, and several classic studies are as follows [3].

Cheng et al. studied UPMSP with degradation and maintenance and proved that the problem could be optimally solved in polynomial time [4]. Avalos-Rosales et al. studied unrelated parallel machines and considered preventive maintenance activities and setup times by order and by machine [5]. Luo J et al. proposed a predictable scheduling and rescheduling and accounting for machine failures and consistency in unrelated machine environments, where work separations include printed circuit boards (PCB) [6].

Comparatively speaking, the research on UPMSP based on the Spider Monkey Optimization (SMO) algorithm is rare. Aiming at the optimization problem of unrelated parallel machine maintenance and scheduling integration, this paper proposes a hybrid spider monkey algorithm, and compares it with classical algorithms to provide the foundation for solving UPMSP [7].

2. Problem Formulation

The problem studies in this paper can be described as follows: n jobs are to be processed on m unrelated parallel machines; in most situations, we assume m is less than n, and these jobs are non-preemptive and can all be processed at time 0. Maintenance performed on the machine may depend on the state of the machine (e.g., running time). The state of a machine is determined by reliability, which decreases with the cumulative processing time of the workpiece or degradation of the machine. Once the reliability of the machine falls below the threshold r_{th}, PM must be implemented. The reliability of the machine does not change during operation.

Using the three-field notation $\alpha \mid \beta \mid \gamma$ for describing scheduling problems, we denote our problem by $\mathrm{Rm/nr, VPM/C_{max}}$, where nr denotes those jobs are non-resumable; "VPM" denotes variable PM; the objective is to minimize the maximum completion time. The decision is to determine the allocation and sequence of n jobs on m machines and the maintenance time of the machines. Since problem $\mathrm{Rm//C_{max}}$ has been proved to be an NP-Hard problem, it can be concluded that problem $\mathrm{Rm/nr, VPM/C_{max}}$ is an NP-Hard problem by comparison. Thus, the approximate methods are needed to solve real-size instances.

3. HDSMO Algorithm

3.1. Basic Flow of the HDSMO Algorithm

SMO is a proposed global optimization algorithm; the main feature is that it can improve the ability to search for optimal solutions. However, in the traditional SMO algorithm, the spider monkey individual SM_h completely inherits the old location information of the individual in the updating process, which makes the algorithm lack the local search ability in the late iteration. An HDSMO algorithm considering inertia weight aims at the above problems and shortcomings. n_{llc} and n_{lll} represent the local leader counter and limit, respectively, while n_{glc} and n_{gll} represents the global leader counter and limit. The process of the proposed HDSMO algorithm is shown in Figure 1.

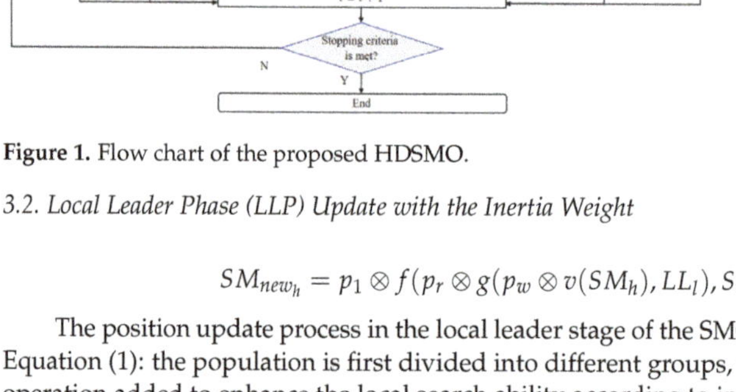

Figure 1. Flow chart of the proposed HDSMO.

3.2. Local Leader Phase (LLP) Update with the Inertia Weight

$$SM_{new_h} = p_1 \otimes f(p_r \otimes g(p_w \otimes v(SM_h), LL_l), SM_r) \tag{1}$$

The position update process in the local leader stage of the SMO algorithm is shown in Equation (1): the population is first divided into different groups, $v(SM_h)$ is the mutation operation added to enhance the local search ability according to inertia weight P_w. For the individuals of the first 50% generation population and the last 50% generation population, the mutation operation methods of reverse order and two-point exchange can be used respectively, which can effectively improve the diversity of the population and further improve the local search ability of the algorithm. The mutation method is shown in Figure 2, where 0 represents the machine, and the remaining numbers represent the job.

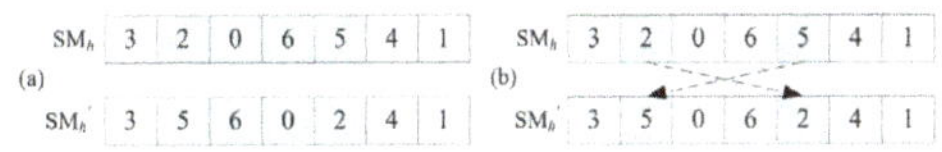

Figure 2. Two mutation operations (**a**) reverse order (**b**) exchange.

$g(p_w \otimes v(SM_h), LL_l)$ and $f(p_r \otimes g(p_w \otimes v(SM_h), LL_l), SM_r)$ represent crossover operations. The mutant individuals cross with LL according to crossover rate P_r, and the generated new individuals cross with random individuals according to crossover rate P_1. In this paper, two crossover methods are designed based on whether there are identical parts between individuals, as shown in Figure 3.

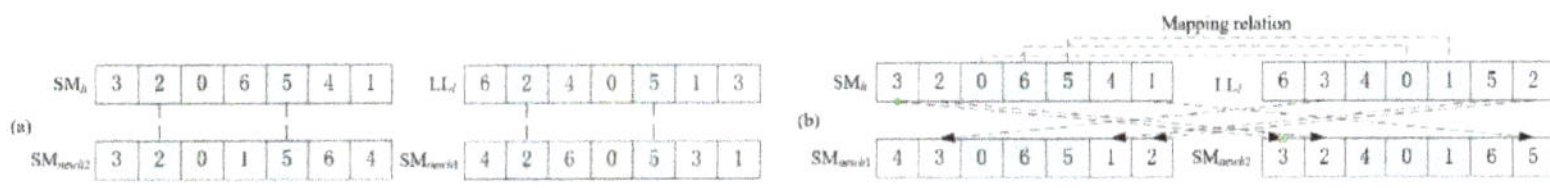

Figure 3. Two kinds of crossover operation (**a**) with the same parts; (**b**) without the same parts.

3.3. Global Leader Phase (GLP) Update with the Inertia Weight

At this phase, individual SM_h mutates according to crossover rate inertia weight P_w, and then crosses with GL according to crossover rate P_r, and the generated new individuals cross with random individuals according to crossover rate P_2. The same method is shown in Section 3.2.

4. Numerical Example and Analysis

4.1. Parameters Setting

The experimental data include the number of machines m, the number of jobs n, the processing time p_{ij}, the PM parameters including the threshold UT, and the maintenance time t^{PM}. For each combination of problem instance size, Generate 10 random problem instances. The instances and the range of experimental parameters are shown in Table 1, the parameters of the GA algorithm and the DSMO algorithm are experimentally analyzed, and the algorithm parameter values under different problem scales are determined as shown in Table 2.

Table 1. Experimental problem scale and parameter range.

Size	m	n	p_{ij}	t^{PM}
Small	2,3	6,8	$U[1, 100]$	10
Medium	4,5	30,50	$U[1, 100]$	20
Large	10,20	100,200	$U[1, 100]$	20

Table 2. Parameter values for algorithms.

Parameter	GA			DSMO			HDSMO		
	Small	Medium	Large	Small	Medium	Large	Small	Medium	Large
N	100	350	500	100	300	450	100	350	200
$Maxt$	200	400	500	100	400	500	100	400	500
p_1	0.4	0.3	0.5	0.3	0.6	0.8	0.2	0.3	0.5
p_2	0.15	0.2	0.2	0.5	0.6	0.8	0.4	0.3	0.5
p_w	/	/	/	/	/	/	0.05	0.25	0.35

4.2. Computational Experiments and Discussion

The computational experiments result for the different algorithms are given in Table 3. Each algorithm calculates the average relative percentage deviation (PD) from the optimal

C_{max} solution, i.e., the value $PD = \frac{C_{max} - C^*_{max}}{C^*_{max}}$. There is also the average computed time in seconds (CT).

Table 3. The performance of the algorithms.

Size	n*m*	GA		DSMO		HDSMO	
		CT	PD	CT	PD	CT	PD
Small	n6m2	1.741	0	0.904	0	3.237	0
	n6m3	1.619	0	1.027	0.027	4.226	0
	n8m2	1.417	0	0.982	0.012	3.45	0
	n8m3	2.239	0.008	1.139	0.031	4.394	0.002
Medium	n30m4	19.499	0.461	43.592	0.436	50.455	0
	n30m5	18.533	0.764	45.173	0.711	52.502	0
	n50m4	25.016	0.899	67.643	0.92	72.982	0.007
	n50m5	24.789	1.137	68.049	1.175	71.131	0
Large	n100m10	121.056	3.843	289.134	3.926	138.101	0
	n100m20	179.374	7.294	329.983	7.1	169.284	0
	n200m10	215.577	4.709	546.221	4.633	247.991	0.002
	n200m20	255.377	10.451	587.604	9.618	265.533	0.005

It can be concluded from Table 3 that HDSMO is superior to DSMO and GA in average relative percentage deviation for three scale problems. However, in terms of computation time, the DSMO algorithm outperforms GA and HDSMO for the small problems, and the needed computation time of HDSMO is decreased with the increase in the problem size. The HDSMO algorithm is a recommended method for solving large and medium-sized problems because it can give approximate optimal solutions in a short computing time.

5. Conclusions

According to the property of the addressed problem and the decision-making method of "job-grouping batch and allocating", a hybrid discrete SMO algorithm is proposed in this paper. Experimental results demonstrate that HDSMO is superior to GA and DSMO in solving quality and effectiveness.

Author Contributions: Conceptualization, Y.C.; methodology, K.K.; software, K.K.; validation, K.K.; formal analysis, K.K.; investigation, K.K.; resources, Y.C.; data—curation, K.K.; writing—original draft preparation, K.K., writing—review and editing, K.K.; visualization, K.K.; supervision, Y.C., project administration, Y.C.; funding acquisition: Y.C. All authors have read and agreed to the published version of the manuscript.

Funding: This research was funded by National Natural Science Foundation of China grant number [No.51705370].

Institutional Review Board Statement: Not applicable.

Informed Consent Statement: Not applicable.

Data Availability Statement: The data that support the findings of this study are available from the corresponding author upon reasonable request.

Conflicts of Interest: The authors declare no conflict of interest.

References

1. Mumtaz, J.; Guan, Z.; Yue, L.; Zhang, L.; He, C. Hybrid Spider Monkey Optimisation Algorithm for Multi-Level Planning and Scheduling Problems of Assembly Lines. *Int. J. Prod. Res.* **2020**, *58*, 6252–6267. [CrossRef]
2. Lei, D.; Liu, M. An Artificial Bee Colony with Division for Distributed Unrelated Parallel Machine Scheduling with Preventive Maintenance. *Comput. Ind. Eng.* **2020**, *141*, 106320. [CrossRef]
3. Mirabedini, S.N.; Iranmanesh, H. A Scheduling Model for Serial Jobs on Parallel Machines with Different Preventive Maintenance (PM). *Int. J. Adv. Manuf. Technol.* **2014**, *70*, 1579–1589. [CrossRef]

4. Cheng, T.C.E.; Hsu, C.J.; Yang, D.L. Unrelated Parallel-Machine Scheduling with Deteriorating Maintenance Activities. *Comput. Ind. Eng.* **2011**, *60*, 602–605. [CrossRef]
5. Avalos-Rosales, O.; Angel-Bello, F.; Álvarez, A.; Cardona-Valdés, Y. Including Preventive Maintenance Activities in an Unrelated Parallel Machine Environment with Dependent Setup Times. *Comput. Ind. Eng.* **2018**, *123*, 364–377. [CrossRef]
6. Luo, J.; Liu, J.; Hu, Y. An MILP Model and a Hybrid Evolutionary Algorithm for Integrated Operation Optimisation of Multi-Head Surface Mounting Machines in PCB Assembly. *Int. J. Prod. Res.* **2017**, *55*, 145–160. [CrossRef]
7. Lei, D.; Yang, H. Scheduling Unrelated Parallel Machines with Preventive Maintenance and Setup Time: Multi-Sub-Colony Artificial Bee Colony. *Appl. Soft Comput.* **2022**, *125*, 109154. [CrossRef]

engineering proceedings

Proceeding Paper

A Multi-Objective Scheduling Optimization Method for PCB Assembly Lines Based on the Improved Spider Monkey Algorithm †

Jingyan Zhong [1], Yarong Chen [1,*] and Jabir Mumtaz [1,2]

1 School of Mechanical and Electrical Engineering, Wenzhou University, Wenzhou 325035, China
2 Department of Mechanical Engineering, Capital University of Science and Technology (CUST), Islamabad 44000, Pakistan
* Correspondence: yarongchen@126.com
† Presented at the 2nd International Conference on Advances in Mechanical Engineering (ICAME-22), Islamabad, Pakistan, 25 August 2022.

Abstract: For assembly lines of circuit printing board (PCBs), this study investigated an integrated scheduling problem of the component allocation problem (CAP) and component placement sequence problem (CPSP) with the fixed interval preventive maintenance to minimize the maximum completion time (C_{max}), the total energy consumption (TEC), and the total maintenance time (TMT) simultaneously. An improved spider monkey optimization (ISMO) algorithm is proposed with selecting the local leader (LL) and the global leader (GL) using a parallel lattice coordinate system. We compared the proposed ISMO algorithm with the classical optimizations, including SMO, NSGA-III, DE, and PSO, by the production data of an enterprise; the results showed that the ISMO algorithm can obtain pareto solutions with better convergence and diversity.

Keywords: ISMO algorithm; multi-objective scheduling optimization; PCB assembly line

Citation: Zhong, J.; Chen, Y.; Mumtaz, J. A Multi-Objective Scheduling Optimization Method for PCB Assembly Lines Based on the Improved Spider Monkey Algorithm. *Eng. Proc.* **2022**, *23*, 15. https://doi.org/10.3390/engproc2022023015

Academic Editors: Mahabat Khan, M. Javed Hyder, Muhammad Irfan and Manzar Masud

Published: 20 September 2022

1. Introduction

With the transformation and upgrading of the manufacturing process to be digital and intelligent, the PCB manufacturing industry has developed rapidly. The scheduling of PCB assembly lines directly affects the benefits of the PCB manufacturing enterprise.

Researchers have studied the scheduling problem of PCB assembly lines. The early research was mainly based on single CAP and CPSP problems. However, the two processes of component allocation and sorting are closely related, and the solution results of CAP have a considerable influence on the solution results of CPSP; thus, the current research is mainly based on the integration of CAP and CPSP. Typical studies on CAP are as follows: Ji et al. studied the CAP problem with minimizing cycle time, and designed a genetic algorithm to solve it [1]; Ho et al. studied the CAP problem with minimizing the total distance and proposed a hybrid genetic algorithm [2]. Typical studies on CPSP problems are as follows: Lin et al. studied the CPSP with maximizing production capacity, minimizing total assembly time, and head movement distance [3]; Grunow et al. proposed a three-stage heuristic method to solve CAP and CPSP problems [4].

Typical studies on integrating CAP and CPSP are as follows: Luo et al. constructed a heuristic algorithm for solving the CPSP problem that combines a genetic algorithm and a tabu search [5]; Mumtaz et al. proposed a hybrid SMO algorithm for the multi-level planning and scheduling of PCB assembly line [6]; Gao et al. proposed a hierarchical multi-objective heuristic algorithm for optimizing PCB assembly [7].

After analyzing the literature, we can find that the current research on the scheduling problem of the PCB assembly line focuses on a single objective; the maintenance of the machine and the energy consumption of the assembly line are rarely considered. Therefore, this study designed an ISMO algorithm for the integrated scheduling problem of CAP and CPSP, aiming to minimize C_{max}, TEC, and TMT at the same time.

2. Problem Formulation

The problem studies in this paper can be described as follows: if CN is the number of PCB orders, then m number of machines for t types of n component patches are required. Each machine will require maintenance at regular intervals to ensure the normal state of machines, i.e., the continuous placement time of the machine cannot exceed the maintenance interval time UT. The focus is to determine the optimum distribution and placement sequence of n components on the m machine to minimize C_{max}, TEC, and TMT simultaneously. The energy consumption of machines is mainly composed of three parts: processing energy consumption, maintenance energy consumption, and idle energy consumption.

Assumptions of this paper include the following: (1) all machines on a PCB assembly line are identical; (2) a machine can only place one component at a time on the PCB; (3) the same type of components can be assigned to multiple machines on the PCB assembly line; (4) there is no priority constraint on component placement; (5) each machine has a feeder, and multiple feed slots can be placed in each feeder, and only one type of component can be placed in each feed slot; (6) the machine cannot stop for maintenance during the placement process.

3. ISMO Algorithm

3.1. Basic Flow of the ISMO Algorithm

SMO is a proposed global optimization algorithm: the main feature is that it can increase the ability to search optimal solutions. The ISMO algorithm is proposed to solve multi-objective problems. The process of the proposed ISMO algorithm is shown in Figure 1.

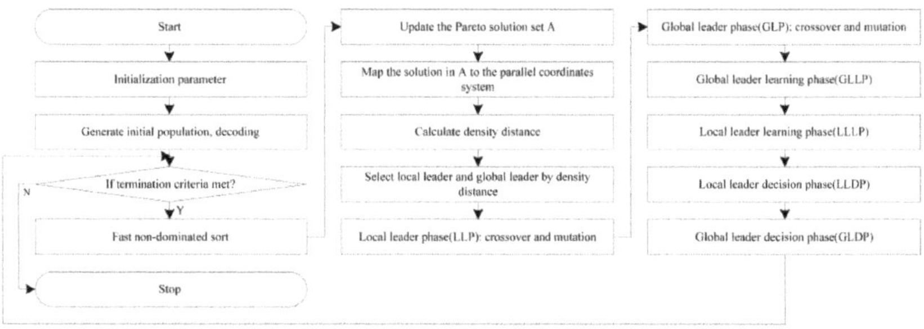

Figure 1. Flow chart of the proposed ISMO.

3.2. Solution Based on Parallel Coordinates

The proposed problem is multi-objective optimization problem; therefore, the Pareto solution set is normalized by parallel coordinates to determine the advantages and disadvantages of each solution. The rth target value corresponding to the $(n^s)^{th}$ solution in the Pareto solution $f_{n^s,r}$ set to a $\left(N^S \times R\right)$ two-dimensional planar mesh (N^S is the number of Pareto solutions, R represents the number of target values), using Equation (1) to obtain the lattice coordinate component $L_{n^s,r}$:

$$L_{n^s,r} = N^S \times \frac{f_{n^s,r} - f_r^{min}}{f_r^{max} - f_r^{min}}, \quad (1)$$

where $f_r^{max} = max f_{n^s,r}$ and $f_r^{min} = min f_{n^s,r}$ are the maximum and minimum values of the rth target of the Pareto solution, respectively. If $f_{n^s,r} = f_r^{min}$, then $L_{n^s,r}$=1.

The density distance $D(SM_u)$ of SM_u can be calculated using Equation (2).

$$D(SM_u) = \sum_{j=1,j\neq i}^{N^S} \frac{PCD(SM_u, SM_v)^2}{N^S}, \quad (2)$$

where SM_v is the individual in the solution set except SM_u, and $PCD(SM_u, SM_v)$ can be calculated using Equation (3).

$$PCD(SM_u, SM_v) = \begin{cases} 0.5, & \forall r, L_{u,r} = L_{v,r} \\ \sum_{r=1}^{R} |L_{u,r} - L_{v,r}|, & \exists r, L_{u,r} \neq L_{v,r} \end{cases}, \quad (3)$$

The individual with the largest density distance in the Pareto solution set is selected as the optimal solution.

3.3. Local Leader Phase (LLP)

In the LLP, the population is divided into groups, and then the individual SM_u is updated in the group as follows: SM_u is crossed with LL first, and the obtained offspring individuals are crossed with the random individual in the group; then, the best individual is selected as the new individual.

The crossover process is illustrated below. The crossover of the components part adopts the two-point crossover method, and the crossover of the machine part adopts the single-point crossover method. The example of crossover is shown in Figure 2.

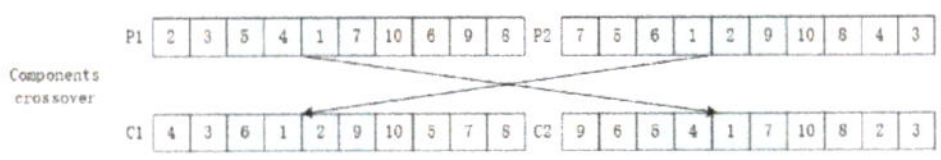

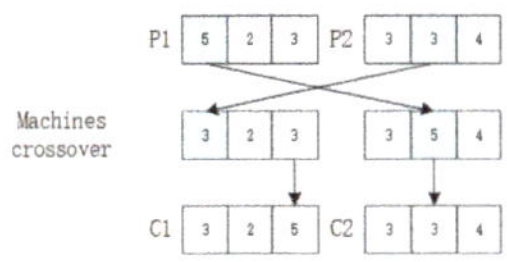

Figure 2. (**a**) Component's crossover; (**b**) machines crossover.

4. Numerical Example and Analysis

4.1. Parameter Settings

In order to evaluate the solution efficiency and quality of the ISMO, SMO, NSGA-III, PSO and DE were selected as the comparison algorithms. Based on the production data of a certain enterprise [8], the experimental problem scale and parameter range are shown in Table 1, and the algorithm parameters are shown in Table 2. n, m, ct, and CN are the number of components, machines, component types, and PCB orders, respectively; UT and MT are the maintenance interval time and the maintenance time; v^m is the velocity of the machine head; d is the distance covered as the machine head moves; and FP, PP, and MP are the idle power, processing power, and maintenance power of the machine, respectively.

Table 1. Experimental problem scale and parameter range.

Parameters		Value		Parameters	Value	Parameters	Value
n	10	20	30	MT	300	FP	U [1, 3]
m	2	3	4	v^m	100	PP	U [1, 9]
ct	4	5	6	d	U [300, 350]	MP	U [1, 99]
CN	10,000	20,000	30,000	UT	14,400		

Table 2. Parameter values for algorithms.

Instance	$m = 2, n = 10$					$m = 3, n = 20$					$m = 4, n = 30$				
Algorithm	**ISMO**	**SMO**	**NSGA-III**	**PSO**	**DE**	**ISMO**	**SMO**	**NSGA-III**	**PSO**	**DE**	**ISMO**	**SMO**	**NSGA-III**	**PSO**	**DE**
popsize	150	100	150	150	150	150	150	150	150	150	150	150	150	150	150
maxgen	150	100	150	200	150	200	150	150	200	200	200	150	200	200	200
p_c	0.7	0.7	0.75	/	/	0.7	0.7	0.75	/	/	0.75	0.75	0.8	/	/
p_m	0.1	0.1	0.2	/	/	0.1	0.1	0.15	/	/	0.15	0.15	0.2	/	/

4.2. Computational Experiments and Discussion

In this study, three problem instances were tested, with each problem instance run 10 times, in order to compare different algorithms in terms of operational efficiency and operational quality. The results are shown in Table 3. CT is the running time of the algorithm; N_d is the average number of Pareto solutions obtained by the algorithm; IGD is inverse generation distance; and NR is non-dominant rate.

Table 3. Performance indicator results for algorithms.

Algorithm	$m = 2, n = 10$				$m = 3, n = 20$				$m = 4, n = 30$			
	CT	N_d	**IGD**	**NR**	**CT**	N_d	**IGD**	**NR**	**CT**	N_d	**IGD**	**NR**
ISMO	13.4198	3.2	1646.5110	0.7861	41.4082	5.5	623,020.0337	0.7231	53.9259	6.8	1,892,077.6829	0.6624
SMO	12.6523	1.5	44,995.7403	0.0111	38.2206	2.4	842,868.3469	0.0000	50.7701	2.5	2,043,876.5376	0.0676
NSGA-III	22.8764	1.3	8391.4363	0.3139	27.2215	1.2	1,595,181.4112	0.2269	45.0490	1.1	5,047,418.3441	0.2699
PSO	12.1930	1.6	22,905.7269	0.0000	14.0099	2.5	1,062,318.0880	0.0125	16.2862	3.0	4,656,473.8169	0.0000
DE	5.1382	1.8	56,670.4167	0.0000	8.3881	2.9	1,075,758.9253	0.0375	10.2588	3.2	4,453,947.1130	0.0000

It can be seen from Table 3 that the CT of DE and PSO algorithms is shorter, and the CT of ISMO, SMO, and NSGA-III algorithm is longer. The DE and PSO are basic algorithms, and the principle and running steps are relatively simple; therefore, the CT is significantly shorter than the other three algorithms. For the number of Pareto solutions obtained by different algorithms, the average order from large to small is: ISMO, DE, PSO, SMO, NSGA-III. When using IGD, the ISMO algorithm is obviously the best, followed by SMO, and the PSO and DE are poor, whereas the NSGA-III has better results in $n = 10$ problem instances, but worse result in the $n = 20$ and $n = 30$ problem instances. When using NR, the ISMO algorithm is obviously the best, followed by NSGA-III, and the SMO, PSO, and DE are worse. In general, compared with the other four algorithms, the ISMO algorithm has obvious advantages, but its advantages decrease with the increase in scale.

5. Conclusions

For the integrated CAP and CPSP scheduling problem considering machine preventive maintenance, this paper has proposed an ISMO algorithm based on the parallel lattice coordinate, which minimizes C_{max}, TEC, and TMT at the same time. The experimental results show that the ISMO algorithm can effectively increase the diversity and convergence of the obtained solutions.

Author Contributions: Conceptualization, Y.C.; methodology, Y.C., J.M., and J.Z.; software, Y.C.; validation, Y.C.; formal analysis, J.Z.; investigation, J.Z.; resources, J.M.; data curation, J.Z.; writing—original draft preparation, J.Z.; writing—review and editing, Y.C.; visualization, J.Z.; supervision, Y.C.; project administration, Y.C.; funding acquisition, Y.C. All authors have read and agreed to the published version of the manuscript.

Funding: This research was funded by National Natural Science Foundation of China grant number [No.51705370].

Institutional Review Board Statement: Not applicable.

Informed Consent Statement: Not applicable.

Data Availability Statement: The data that support the findings of this study are available from the corresponding author upon reasonable request.

Acknowledgments: All individuals included in this section have consented to the acknowledgement.

Conflicts of Interest: The authors declare no conflict of interest.

References

1. Ji, P.; Sze, M.T.; Lee, W.B. Genetic Algorithm of Determining Cycle Time for Printed Circuit Board Assembly Lines. *Eur. J. Oper. Res.* **2001**, *128*, 175–184. [CrossRef]
2. Ho, W.; Ji, P. A Genetic Algorithm to Optimise the Component Placement Process in PCB Assembly. *Int. J. Adv. Manuf. Technol.* **2005**, *26*, 1397–1401. [CrossRef]
3. Lin, C.J.; Huang, M.L. Modified Artificial Bee Colony Algorithm for Scheduling Optimization for Printed Circuit Board Production. *J. Manuf. Syst.* **2017**, *44*, 1–11. [CrossRef]
4. Grunow, M.; Günther, H.O.; Schleusener, M.; Yilmaz, I.O. Operations Planning for Collect-and-Place Machines in PCB Assembly. *Comput. Ind. Eng.* **2004**, *47*, 409–429. [CrossRef]
5. Luo, J.; Liu, J.; Hu, Y. An MILP Model and a Hybrid Evolutionary Algorithm for Integrated Operation Optimisation of Multi-Head Surface Mounting Machines in PCB Assembly. *Int. J. Prod. Res.* **2017**, *55*, 145–160. [CrossRef]
6. Mumtaz, J.; Guan, Z.; Yue, L.; Wang, Z.; Ullah, S.; Rauf, M. Multi-Level Planning and Scheduling for Parallel PCB Assembly Lines Using Hybrid Spider Monkey Optimization Approach. *IEEE Access* **2019**, *7*, 18685–18700. [CrossRef]
7. Gao, H.; Li, Z.; Yu, X.; Qiu, J. Hierarchical Multiobjective Heuristic for PCB Assembly Optimization in a Beam-Head Surface Mounter. *IEEE Trans. Cybern.* **2021**, *52*, 1–14. [CrossRef] [PubMed]
8. Cao, J.; Guan, Z.; Yue, L.; Ullah, S.; Sherwani, R.A.K. A Bottleneck Degree-Based Migrating Birds Optimization Algorithm for the PCB Production Scheduling. *IEEE Access* **2020**, *8*, 209579–209593. [CrossRef]

engineering proceedings

MDPI

Proceeding Paper

Minimizing a Just-In-Time Objective on a Single-Batch-Processing Machine Using a Hybrid Differential Evolution Algorithm †

Chen Wang, Yarong Chen *, Fuh-Der Chou and Shenquan Huang

School of Mechanical and Electrical Engineering, Wenzhou University, Wenzhou 325035, China
* Correspondence: yarongchen@126.com
† Presented at the 2nd International Conference on Advances in Mechanical Engineering (ICAME-22), Islamabad, Pakistan, 25 August 2022.

Abstract: A hybrid differential evolution (HDE) algorithm that minimizes the total earliness and tardiness time (ET), a just-in-time objective, is studied for a single-batch-processing machine (SBPM) scheduling problem with different processing times, release times, sizes and delivery times. A hybrid differential evolution algorithm with an integrated Tabu search (TS) algorithm is proposed to improve the capacity of the neighborhood search of a differential evolution algorithm (DE). The experimental results show that the proposed HDE can obtain better an objective value than the basic DE can.

Keywords: differential evolution algorithm; Tabu search algorithm; single-batch-processing machine scheduling problem

Citation: Wang, C.; Chen, Y.; Chou, F.-D.; Huang, S. Minimizing a Just-In-Time Objective on a Single-Batch-Processing Machine Using a Hybrid Differential Evolution Algorithm. *Eng. Proc.* **2022**, *23*, 4. https://doi.org/10.3390/engproc2022023004

Academic Editors: Mahabat Khan, M. Javed Hyder, Muhammad Irfan and Manzar Masud

Published: 20 September 2022

1. Introduction

The problem of batch processing machine (BPM) scheduling is a significant branch of the production scheduling problem. The scheduling problem of burn-in operations in semiconductor manufacturing is a BPM scheduling problem. Therefore, it is important to carry out research on BPM scheduling in the context of intelligent manufacturing.

Many research studies have been conducted in the literature. Donbson and Nambimadom studied the SBPM scheduling problem to optimize the mean weighted flow time with different job sizes [1]. Chou et al. designed an improved genetic algorithm to solve the SBPM scheduling problem with different sizes and delivery times of jobs [2]. Zhou improved the DE by selecting the mutation operator adaptively for the same problem that is mentioned in the literature to minimize the makespan [3]. In order to optimize the maximum lateness for the SBPM problem with non-identical job sizes and release dates, Zhou presented a modified particle swarm optimization (MPSO) algorithm [4]. Parsa et al. designed a dynamic programming algorithm to optimize the BPM scheduling for the JIT objective with a common due date, and different processing times and sizes of jobs [5]. Zhang et al. designed SLTS heuristic rules for the SBPM scheduling problem with different processing times and due dates, which can effectively minimize the ET [6]. Zhang et al. designed AABF and ITSLS rules to optimize the total weighted earliness and tardiness [7].

This article researches an SBPM problem to minimize the ET objective, and a hybrid differential evolution algorithm is designed to minimize the ET.

2. Problem Description

The problem can be described like this: n jobs will be processed on a SBPM. First, the jobs will be assigned to a batch and one job can only be assigned to a batch once. The total size of the jobs in a batch cannot be greater than the constraint of the machine S. Then, the processing sequence of the batches is determined to optimize the ET.

Some assumptions need to be considered: (1) p_j, r_j, s_j and d_j, respectively, represent the processing time, release time, size and delivery time of job j. (2) The process cannot be interrupted, even if there are still some jobs in the batch have not been finished. (3) The largest release time of the jobs in the batch determines the release time of a batch. (4) The longest processing time of the jobs in the batch determines the processing time of a batch. Using the three-field notation $\alpha|\beta|\gamma$ which was introduced for describing scheduling problems, we denote our problem by $1|p_j, r_j, d_j, s_j, S|ET_{min}$.

3. Description of DE and HDE

As a frequently used evolutionary algorithm, DE is widely used. DE includes population initialization, mutation, crossover, selection operations, and it can quickly explore solution space.

3.1. Initialization

The initial population of DE is a P × n set that is constructed by P individuals. P denotes the population size. Each individual has n elements, and n denotes the number of jobs that there are. Therefore, the initial population can be denoted by $x_i^t = (x_{i,1}^t, x_{i,2}^t, \cdots, x_{i,n}^t)$, which represents the ith individual in a generation $t(i$ = 1, ..., P). In this article, a job permutation list is obtained by a ranked-order-value (ROV) rule, which is dependent on the random key representation. As Table 1 shows, the random key of the jobs is {0.3, 0.6, 0.1, 0.8, 0.4, 0.9}, and we can get the permutation of job X = {3, 1, 5, 2, 4, 6} by using ROV. Then, a first-fit (FF) rule is employed to group the jobs that are sorted into batches. We assume that the capacity of the machine is $S = 10$, and the two batches.

Table 1. Job permutation obtained by ROV.

Jobs	1	2	3	4	5	6
Random key	0.3	0.6	0.1	0.8	0.4	0.9
Job permutation	3	1	5	2	4	6
Job size	1	8	5	1	3	1

B $= \{b_1, b_2\} = \{\{3,1,2\}, \{5,4,6\}\}$ will be obtained. The individual X_i^t, i = 1, ..., P is randomly generated. Each random key value of the individual is generated by Formula (1)

$$x_{i,j}^0 = x_d + random \times (x_u - x_d), \ j = 1, \cdots, n \tag{1}$$

x_d = −1, x_u = 1, *random* is a random value, *random* ∈ [0, 1].

3.2. Mutation

In this step, this article applies *DE/rand/2* to generate the mutant individuals v_i^t. The mutation operator can be described as Equation (2):

$$V_i^t = X_a^t + F \times (X_b^t - X_c^t) + F \times (X_d^t - X_e^t) \tag{2}$$

where a, b, c, d, e are not equal to each other and are respectively generated in the set {1, ..., P}, and F is randomly chosen in the range [0, 1].

3.3. Crossover

A trial individual U_i^t will be produced in the crossover operator, j_{random} is a random natural number $j_{random} \in \{1, \ldots, n\}$. The CR determines whether an element of V_i^t can be a part of U_i^t, and the CR is randomly generated in the range [0, 1]. The crossover operator can be described as Equation (3):

$$u_{i,j}^t = \begin{cases} v_{i,j}^t & if\ random \le CR\ or\ j = j_{random} \\ x_{i,j}^t & otherwise \end{cases} \quad j = 1, \cdots, n \tag{3}$$

3.4. Selection

In this step, we demonstrate how to compare the fitness of U_i^t with V_i^t and select the better one which will be used as a new individual of the next population. f represents the fitness value.

$$x_i^{t+1} = \begin{cases} U_i^t & if\ f(U_i^t) \leq f(X_i^t) \\ X_i^t & otherwise \end{cases} \tag{4}$$

Equation (4) describes how to select an individual to become a new part of the next population.

3.5. Combination of Two Algorithms

The DE has a weak local search capacity. Therefore, a local search strategy is embedded into the DE to improve the local search capacity of DE, as it is proposed. The TS algorithm is a common local search algorithm. The TS sets a Tabu list to forbid some repeated operations and uses aspiration criterion to assoil a good solution status. After the selection operator of DE, the next step is choosing the best individual of this population as a candidate solution to perform the Tabu operation. Introduced next is 2-opt which produces a new candidate solution. As shown in Figure 1, by exchanging two and four, we can obtain a new solution.

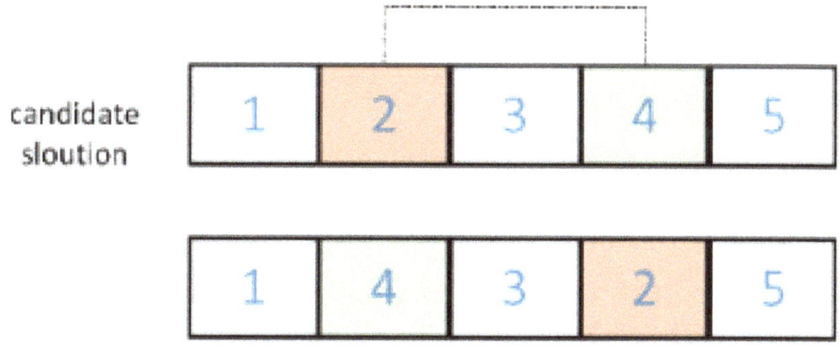

Figure 1. The process of 2-opt.

3.6. The Process to Sort the Batches

The batches are sorted according to the median of the job due date in the batch, and the steps are as follows:

1. They are given a job permutation, and the jobs are assigned to a batch by the FF rule, and b batches are obtained;
2. The median of the jobs due dates $d_{B_i}^{median}$, (i = 1, … , b) in the batch are found, and they are sorted in to an increasing order which decides the order of the batches that are being processed: $B_1, B_2, \ldots, B_b$;
3. The start time of the batch $S_{B_i} = max\{r_{B_i}, C_{B_{i-1}}\}$ is, when i = 1, C_{B_0}= 0. The completion time of the batch is checked, to see if $C_{B_i} < d_{B_i}^{median}$, B_i is shifted from the left to the right until $C_{B_i} = d_{B_i}^{median}$; if the situation is otherwise, then nothing needs to be changed. The operation is repeated until all the batches are scheduled.

4. Computational Experiments

4.1. Data Generation

Extensive computational experiments are done. Let n = 10, 12, 15, 20, 40, 60, 80, 100 for these. The processing time p_j, the release time r_j and the size s_j of job were, respectively, discrete distributions in the range of [1, 100], [0, 50] and, [1, 10]. The due date of the job d_j satisfies the following formulae $d_j = p_j + U[d_{min}, d_{min} + \beta \times P]/2$ and $d_{min} = \max(0, P \times (\alpha - \beta/2))$, $P = \sum_{j=1}^{n} p_j$, $\alpha = 1$, $\beta = 0.5$. The machine's capacity is S = 30.

4.2. Results Analysis

To compare the quality of the DE and HDE, each scale data were randomly produced in 10 instances. Table 2 shows the average values of the completion time (CT) and the ET that were obtained by testing every scale of the jobs.

Table 2. Results that were obtained by different algorithms.

n	DE		HDE		Gap (%)
	CT	ET	CT	ET	
10	2.24	224.3	9.04	224.3	0
12	2.72	309.9	20.96	306.2	−0.01
15	13.39	384.5	27.30	378.2	−0.02
20	52.81	617.8	106.42	585.4	−0.05
40	110.12	2035.5	207.26	1434.9	−0.30
60	168.83	5744.6	570.46	2807.1	−0.51
80	222.60	10966.4	803.57	4304.3	−0.61
100	274.67	22962.7	925.48	6649	−0.71

The sixth Column of Table 2 shows the gap between the DE and HDE, $Gap = (ET_{HDE} - ET_{DE})/ET_{DE} \times 100\%$. According to Table 2, the ET that was obtained by HDE is close to the DE when $n \leq 0$, because two kinds of algorithm can explore the solution space completely. With the increase of job scale, the ET that can be obtained by HDE is much better than the DE when $n \geq 40$, which indicates that the exploitation capacity of HDE is obviously better than DE.

5. Conclusions

In this paper, a SBPM scheduling problem with non-identical processing times, release times, due dates and arbitrary job sizes to minimize the ET was addressed. An HDE was proposed according to the characteristics of the problem. Compared to the basic DE, HDE can achieve a better solution, especially in situations where there are more than 60 jobs. Future research may consider the parallel-batch processing machine problem. At the same time, many objectives will be considered to be closer to the reality.

Author Contributions: Conceptualization, C.W. and Y.C.; methodology, C.W. and S.H.; software, C.W. and F.-D.C.; validation, C.W. and F.-D.C.; formal analysis, C.W. and Y.C.; investigation, F.-D.C. and S.H.; data curation, Y.C.; writing—original draft preparation, C.W.; writing—review and editing, Y.C. and S.H.; supervision, S.H.; project administration, C.W. All authors have read and agreed to the published version of the manuscript.

Funding: This research was funded by National Natural Science Foundation of China grant number [No. 51705370].

Institutional Review Board Statement: Not applicable.

Informed Consent Statement: Not applicable.

Data Availability Statement: The data that support the findings of this study are available from the corresponding author upon reasonable request.

Acknowledgments: All individuals included in this section have consented to the acknowledgement.

Conflicts of Interest: The authors declare no conflict of interest.

References

1. Dobson, G.; Nambimadom, R.S. The Batch Loading and Scheduling Problem. *Oper. Res.* **2001**, *49*, 52–65. [CrossRef]
2. Der Chou, F. A Joint GA+DP Approach for Single Burn-in Oven Scheduling Problems with Makespan Criterion. *Int. J. Adv. Manuf. Technol.* **2007**, *35*, 587–595. [CrossRef]

3. Zhou, S.; Xing, L.; Zheng, X.; Du, N.; Wang, L.; Zhang, Q. A Self-Adaptive Differential Evolution Algorithm for Scheduling a Single Batch-Processing Machine with Arbitrary Job Sizes and Release Times. *IEEE Trans. Cybern.* **2021**, *51*, 1430–1442. [CrossRef] [PubMed]
4. Zhou, H.; Pang, J.; Chen, P.K.; Chou, F. Der A Modified Particle Swarm Optimization Algorithm for a Batch-Processing Machine Scheduling Problem with Arbitrary Release Times and Non-Identical Job Sizes. *Comput. Ind. Eng.* **2018**, *123*, 67–81. [CrossRef]
5. Parsa, N.R.; Karimi, B.; Husseini, S.M.M. Exact and Heuristic Algorithms for the Just-in-Time Scheduling Problem in a Batch Processing System. *Comput. Oper. Res.* **2017**, *80*, 173–183. [CrossRef]
6. Zhang, H.; Wu, F.; Yang, Z. Hybrid Approach for a Single-Batch-Processing Machine Scheduling Problem with a Just-in-Time Objective and Consideration of Non-Identical Due Dates of Jobs. *Comput. Oper. Res.* **2021**, *128*, 105194. [CrossRef]
7. Zhang, H.; Yang, Y.; Wu, F. Just-in-Time Single-Batch-Processing Machine Scheduling. *Comput. Oper. Res.* **2022**, *140*, 105675. [CrossRef]

engineering proceedings

Proceeding Paper

Investigation and Prioritization of Manpower Activities and Strategic Human Resource Management Factors in Human Resource Information System †

Laraib Habib * and Muhammad Sajid

Department of Industrial Engineering, University of Engineering and Technology Taxila, Punjab 47050, Pakistan
* Correspondence: habib.laraib@gmail.com

† Presented at the 2nd International Conference on Advances in Mechanical Engineering (ICAME-22), Islamabad, Pakistan, 25 August 2022.

Abstract: Human Resource(HR) is one of the main departments in an organization that foresees all the manpower of a company. HR trends are changing rapidly. One such advancement in the field of HR is the human resource information system (HRIS). This paper aims to investigate and prioritize the factors of manpower activities and strategic HRM in HRIS. For this purpose, analytic hierarchy process (AHP) has been used. The result showed that HRs use HRIS for basic purposes such as training, hiring, and forming HR policies. This paper provides an insight into the investigation of the most important factors while implementing HRIS in an organization.

Keywords: human resource information system; manpower activities; strategic human resource management; analytic hierarchy process

Citation: Habib, L.; Sajid, M. Investigation and Prioritization of Manpower Activities and Strategic Human Resource Management Factors in Human Resource Information System. *Eng. Proc.* **2022**, *23*, 8. https://doi.org/10.3390/engproc2022023008

Academic Editors: Mahabat Khan, M. Javed Hyder, Muhammad Irfan and Manzar Masud

Published: 20 September 2022

1. Introduction

Technology is evolving rapidly with time. So, businesses and other organizations must adapt to new trends to stay in the competition. Technology helps businesses in managing and retrieving their database, keeping records, increasing office productivity, and much more [1]. Mid- to high-level companies use the human resource information system (HRIS) to achieve all these goals with just a few clicks. Some research regarding the impacts of HRIS on human resources has already been conducted. Hendrickson [2] has defined HRIS as an integrated system that helps the organization gather, store, analyze and process information regarding its human resource. Mark Feffer [3] recognized human resource information system as a trend that will only grow.

It is new terminology, and many are unaware of it, especially in Pakistan. This paper helps to identify and investigate what Pakistani engineering firms use HRIS for and how they prioritize the manpower activities and strategic HRM factors in HRIS. These preferences have been obtained using the multicriteria decision-making (MCDM) technique called analytic hierarchy process (AHP).

2. Literature Review

Human resource information system has a vast impact on various aspects of an organization from hiring to career management to strategic human resource management. Researchers have been finding ways to define HRIS and its impacts on organizations. Lippert and Swiercz [4] focused on the user's trust in the technology. The greater the trust, the easier it will be to deploy the HRIS. Obeidat [5] talked about HRIS and its relationship with human resource functionalities. He described HRIS as a system that is comprised of strategic tools and planners that improve the future forecast of the supply and demand of the workforce and enhance workers' performances.

Regarding HRIS and manpower activities, Fernandez [6] studied the impact of human resource information system on the recruitment of new employees. With the increase in competition, it is very important for companies to hire better people for the job and to ease the pain of hiring and recruiting. Khera and Gulati [7] did not limit her study to recruitment like Fernandez. She tried to observe the impact of the HRIS on other human resources activities such as training, HR planning, etc. The results after data analysis showed that HRIS has many benefits but one of the best benefits is its capability of storing and analyzing employees' data.

HRIS has been changing the face of traditional HRM and is converting it into more of an advanced HRM with the use of technology; it is known as strategic human resource management (SHRM). Nagendra and Deshpande [8] focused on the strategic side of human resource management. According to this study, aligning IT with HR strategies would help the organization improve its HR tasks. Cheema [9] focused on the decision-making side of the SHRM System.

Vargas [10] explained the analytic hierarchy process(AHP) and its applications in different fields. This study described the two processes that are involved in AHP. The first process includes hierarchic design of pairwise comparisons and the second process deals with evaluation of that comparison. Esangbedo et al. [11] mostly focused on the use of multi criteria decision making techniques in human resource information system.

3. Research Methodology

This study aims to investigate and find the preferences of factors regarding manpower activities and SHRM in HRIS. For this purpose, AHP was used. The first step was to identify all the significant factors in manpower activities (hiring, training, job satisfaction, promotion/demotion, wages, incentives, employees' health, employees safety, communication, staffing, and leaves) and SHRM (career management, demand/supply, HR policies, employee's rights and policies, HR development, job analysis, work–study method, ratio trend analysis, risk management, decision making, business process re-engineering (BPR), workplace learning, HR planning, e-learning, globalization, and offshoring). The next step was to make questionnaires (for manpower activities and SHRM) that would help to assign weights in a pairwise comparison. The weights were assigned using a preference table. The questionnaire was filled out by HR experts. After obtaining the response, the third step was to put in all the weights of all the factors and to find the prioritized factors using AHP (by using expert choice). At first, manpower activities factors were put into the software. Weights were assigned to the factors and the results were obtained. After that, the same steps were followed for SHRM factors.

4. Results and Findings

After putting the weights for manpower activities in expert choice, the comparison matrix obtained is shown in Figure 1.

Compare the relative importance with respect to: Goal: To Find which manpower factors expert prefer for HRIS

	Hiring	Training	Job Satisfa	Promotion/	Wages	Incentives	Employee	Employee'	Communic	Staffing	Leaves
Hiring		4.0	1.0	5.0	2.0	4.0	2.0	6.0	1.0	5.0	3.0
Training			3.0	1.0	2.0	3.0	2.0	3.0	2.0	2.0	1.0
Job Satisfaction				2.0	3.0	2.0	1.0	4.0	4.0	3.0	2.0
Promotion/Demotion					4.0	3.0	2.0	2.0	1.0	3.0	1.0
Wages						1.0	2.0	2.0	3.0	1.0	1.0
Incentives							1.0	2.0	3.0	2.0	2.0
Employee Health								2.0	4.0	5.0	2.0
Employee's Safety									3.0	2.0	2.0
Communication										3.0	2.0
Staffing											2.0
Leaves	Incon: 0.08										

Figure 1. Pairwise comparison matrix obtained from the software "Expert Choice" for manpower activities.

From Figure 2, it can be seen that for manpower activities, the preferred factors include hiring (0.208), and communication (0.166), while staffing showed the lowest value of 0.032.

Figure 2. Preference table for manpower activities using expert choice.

The next step was to put in the alternatives for SHRM and assign weights to them using the software. The pairwise comparison matrix for SHRM is shown in Figure 3.

Compare the relative importance with respect to: Goal: To find which SHRM factors expert prefer for HRIS

	Career Mar	Demand ar	HR Policie	Employee'	HR Develo	Job Analys	Work-Study	Ratio Tren	Risk Mana	Decision M	BPR	Workplace	HR Plannir	E-learning	Globalizati	Offshoring
Career Management		1.0	1.0	1.0	1.0	1.0	4.0	4.0	1.0	1.0	1.0	6.0	5.0	5.0	1.0	1.0
Demand and Supply			3.0	3.0	3.0	1.0	2.0	1.0	1.0	3.0	1.0	3.0	1.0	1.0	1.0	1.0
HR Policies				5.0	3.0	5.0	1.0	3.0	3.0	2.0	1.0	1.0	3.0	2.0	5.0	4.0
Employee's rights and policies					1.0	5.0	3.0	6.0	4.0	5.0	4.0	7.0	5.0	3.0	2.0	3.0
HR Development						3.0	2.0	4.0	3.0	2.0	4.0	4.0	2.0	5.0	1.0	3.0
Job Analysis							3.0	4.0	5.0	4.0	3.0	4.0	5.0	3.0	1.0	1.0
Work-Study Method								1.0	1.0	1.0	1.0	1.0	4.0	2.0	2.0	1.0
Ratio Trend Analysis									2.0	1.0	2.0	1.0	1.0	1.0	1.0	1.0
Risk Management										1.0	1.0	1.0	6.0	3.0	2.0	1.0
Decision Making											2.0	2.0	4.0	3.0	1.0	1.0
BPR												1.0	1.0	1.0	1.0	1.0
Workplace Learning													1.0	1.0	1.0	4.0
HR Planning														4.0	5.0	5.0
E-learning															4.0	1.0
Globalization																1.0
Offshoring																

Figure 3. Pairwise comparison matrix obtained from the software "Expert Choice" for SHRM.

For SHRM, the results after assigning the weights to all the factors showed the following priority: HR policies, job analysis, HR development, career management, HR planning, employees, rights and policies, decision making, and so on. This order shows that for SHRM, HRs prefer HRIS for forming and managing HR policies (0.120), keeping a record of employees and job analysis (0.110), developing HR department (0.10), and employees' career management (0.094). Further results are presented in Figure 4.

Figure 4. Preference table for SHRM using expert choice.

5. Conclusions

The main purpose of this study was to identify the priorities of human resource regarding manpower activities and strategic human resource information system while using HRIS. In this research work, analytical hierarchy process was used to prioritize the factors.

Hiring, training, promotion/demotion, HR development, policy making, and job analysis were found the most important factors while studying the manpower activities and strategic human resource management. The major limitation of this study includes the use of traditional (outdated) HR systems in engineering firms. Most of the HR experts were found unaware of the use HRIS and its working which opens the way for HR researchers to investigate this limitation in future.

Author Contributions: Conceptualization, L.H. and M.S.; methodology, L.H. and M.S.; software, L.H.; validation, L.H. and M.S.; formal analysis, L.H. and M.S.; investigation, L.H. and M.S.; data curation, L.H. writing—original draft preparation, L.H.; writing—review and editing, M.S.; supervision, M.S.; project administration, M.S. All authors have read and agreed to the published version of the manuscript.

Funding: This research received no external funding.

Institutional Review Board Statement: Not applicable.

Informed Consent Statement: Informed consent was obtained from all subjects involved in the study.

Data Availability Statement: The questionnaire that was distributed among HRs can be accessed using the following link: https://drive.google.com/file/d/1098dGJmRfWEieQIbSWYBo3mB9RmUooFk/view?usp=sharing (accessed on 16 August 2022).

Conflicts of Interest: The authors declare no conflict of interest.

References

1. Cascio, W.F.; Montealegre, R. How Technology Is Changing Work and Organizations. *Annu. Rev. Organ. Psychol. Organ. Behav.* **2016**, *3*, 349–375. [CrossRef]
2. Hendrickson, A.R. Human Resource Information Systems: Backbone Technology of Contemporary Human Resources. *J. Labor Res.* **2003**, *24*, 381–394. [CrossRef]
3. Feffer, M. The Impact of Technology on HR: The Good, Bad and Daunting. Available online: https://www.techtarget.com/searchhrsoftware/feature/The-impact-of-technology-on-HR-The-good-bad-and-daunting (accessed on 30 June 2022).
4. Lippert, S.K.; Swiercz, P.M. Human Resource Information Systems (HRIS) and Technology Trust. *J. Inf. Sci.* **2005**, *31*, 340–353. [CrossRef]
5. Obeidat, B.Y. The Relationship between Human Resource Information System (HRIS) Functions and Human Resource Management (HRM) Functionalities. *J. Manag. Res.* **2012**, *4*, 192–211. [CrossRef]
6. Fernandez, J.A.; Valdes-Conca, J. Use of HRIS in the Recruitment Process. In Proceedings of the European and Mediterranean Conference on Information Systems, Costa Blanca, Spain, 6–7 July 2006; pp. 1–7.
7. Khera, S.N.; Gulati, K. Human Resource Information System and Its Impact on Human Resource Planning: A Perceptual Analysis of Information Technology Companies. *IOSR J. Bus. Manag.* **2012**, *3*, 6–13. [CrossRef]
8. Nagendra, A.; Deshpande, M. Human Resource Information Systems (HRIS) in HR Planning and Development in Mid to Large Sized Organizations. *Procedia Soc. Behav. Sci.* **2014**, *133*, 61–67. [CrossRef]
9. Cheema, S.; Javed, F.; Akram, A.; Pasha, A.T. The Impact of HRIS and HPWS on Employee Performance: Surveyed from SMEs of Pakistan Manufacturing Firms. *Organ. Behav.* **2016**, *91*, 38270–38276.
10. Vargas, L.G. An Overview of the Analytic Hierarchy Process and Its Applications. *Eur. J. Oper. Res.* **1990**, *48*, 2–8. [CrossRef]
11. Esangbedo, M.O.; Bai, S.; Mirjalili, S.; Wang, Z. Evaluation of Human Resource Information Systems Using Grey Ordinal Pairwise Comparison MCDM Methods. *Expert Syst. Appl.* **2021**, *182*, 115151. [CrossRef]

engineering proceedings

MDPI

Proceeding Paper

Reinforcement of a Gas Transmission Network †

Muhammad Naveed * **and Saif Ullah**

Department of Industrial Engineering, University of Engineering and Technology, Taxila 47080, Pakistan
* Correspondence: mnaveed48@gmail.com

† Presented at the 2nd International Conference on Advances in Mechanical Engineering (ICAME-22), Islamabad, Pakistan, 25 August 2022.

Abstract: There have been many studies performed on the optimal design and expansion of a natural gas transmission network; however, very few works have addressed the problem of the reinforcement of an existing natural gas transmission network with limited application. This study is focused on the reinforcement of an existing natural gas transmission network with the aim to minimize investment cost. The compressor stations have been assumed operational in either direction. A mathematical model was developed for the problem, which is non-convex mixed integer nonlinear programming (MINLP) in nature; therefore, a convex relaxation was formulated to solve the problem easily in General Algebraic Modeling System (GAMS, GAMS Development Corp., Fairfax, VA, USA) using DICOPT and CONOPT solvers. The model was applied to a small transmission network for validation and the results proved its efficiency.

Keywords: natural gas; transmission network; reinforcement planning; optimal expansion; mixed integer nonlinear programming

check for updates

Citation: Naveed, M.; Ullah, S. Reinforcement of a Gas Transmission Network. *Eng. Proc.* **2022**, *23*, 30. https://doi.org/10.3390/engproc2022023030

Academic Editors: Mahabat Khan, M. Javed Hyder, Muhammad Irfan and Manzar Masud

Published: 23 September 2022

1. Introduction

The natural gas transmission network is the most critical part of a gas supply chain network. The different gas supply sources including the indigenous gas-producing sources, Liquefied Natural Gas (LNG) terminals, underground storage, and international pipeline entry points are connected to a natural gas transmission network at borders. Major industrial consumers such as gas-fired electric power plants, and fertilizer and cement plants are directly connected to a natural gas transmission network. Therefore, its efficient operation and optimal expansion are very critical to ensuring an uninterrupted supply of natural gas as well as minimal cost [1]. Natural gas transmission networks are high-pressure networks spanning entire countries involving large distances. The natural gas transmission networks expanded with time due to the addition of the new demand centers and the enhanced load at the existing ones. The flow dynamics of natural gas through a pipeline are very complex due to the nonlinear relationship between the natural gas pressure and flow [2]. Further, the combinatorial nature of pipeline diameters makes the problem mixed-integer in nature [3]. These characteristics make the problem an Mixed Integer Nonlinear Programming (MINLP), which further needs reformulations and relaxation for efficient results. In literature, the unidirectional operation of the compressor stations has been considered, which makes the compressor station nonoperational in the reverse direction, hence, it has no role in the capacity of the cyclic network in the reverse direction. In this study, it is assumed that they can operate in either direction to fully utilize their capacity, hence, the network. The objective of this study is to propose a mathematical formulation for the optimal expansion of a gas transmission network while considering the reinforcement of the existing network with the bidirectional flow of the compressor stations. The substantial costs can be saved by even minor improvements in design and operation conditions. The total cost can be minimized by optimizing system variables while ensuring that all network constraints are satisfied [4]. A typical natural gas transmission network is shown in Figure 1.

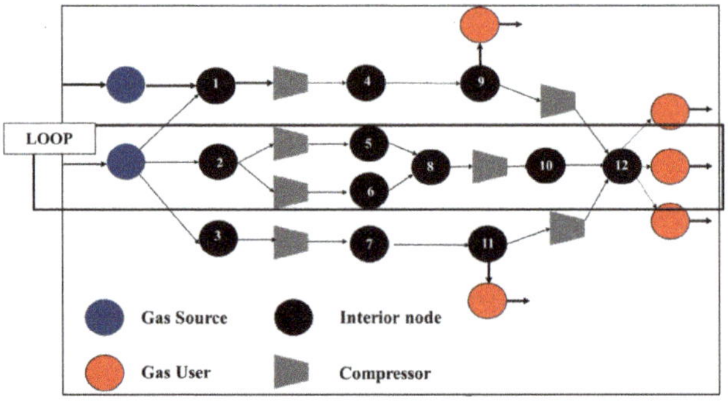

Figure 1. A Cyclic Gas Network.

2. Problem Formulation and Mathematical Modeling

We are dealing with the reinforcement of an existing gas transmission network with the following assumptions: The network under study is a cyclic gas network and is in a steady-state condition. The gas composition and temperature are assumed as constant over time and space. The variations in elevation of the network are neglected and it is considered horizontal. The demand at each delivery node must be fully satisfied, i.e., no shortages are allowed.

Let,

i, index of the nodes; a, index of the arcs;
N, the set of nodes;
N_s, the set of supply nodes;
N_t, the set of transit nodes;
N_d, the set of demand nodes;
A, the set of arcs;
A_{pipe}, the set of pipe arcs;
A_{comp}, the set of compressor arcs;
D, the set of commercially available diameters of pipe;
Q_a, the flow rate in a pipe (million standard cubic feet per day);
π_a^i, the inlet square pressure (or inlet head) of the pipe (psi);
π_a^j, the outlet square pressure (or outlet head) of the pipe (psi);
y_a, binary variable showing gas flow direction in a pipe;
k_a, auxiliary variable for a pipe arc;
L, the pipe length (miles);
D_a, the internal diameter of the pipe (inches);
M, nodes-arcs incidence matrix;
b, the vector of supply or demand;
P_{max}, the maximum pressure at a node (psi;
P_{min}, the minimum pressure at a node (psi);
π_i^{max}, the maximum pressure head limit;
π_i^{min}, the minimum pressure head limit;
C_a, the cost of pipe per mile per inch.

2.1. Objective Function

The objective function is to minimize the construction cost of new pipelines, which is given by:

$$\text{Min } Z = \sum_{a \in A_{pipe}} C_a \cdot D_a \cdot L_a \quad (1)$$

2.2. Constraints

There are some limitations posed by the technicalities of a gas transmission network, therefore, the model is subject to the following constraints:

$$\pi_a^i - \pi_a^j = \beta Q_a|Q_a|, \forall\ a\ \varepsilon\ A_{pipe} \tag{2}$$

$$\pi_i^{min} \leq \pi_i \leq \pi_i^{max}, \forall\ i\ \varepsilon\ N \tag{3}$$

$$\tau_a^{min} \leq \tau_a \leq \tau_a^{max}, \forall\ a\ \varepsilon\ A_{comp} \tag{4}$$

$$MQ = b \tag{5}$$

where β is a constant in the pressure loss equation; whereas, the modulus of Q_a shows the direction of gas flow in a pipe arc. Further, Equation (5) summarizes flow balance equations at each node.

3. Solution Methodology

The relation between the gas flow and pressure makes the problem non-convex in nature, which is difficult to solve; therefore, the convex relaxation is applied to the constraint (2) of the model, then, solved in GAMS.

3.1. Convex Relaxation

The model with the convex relaxation of (2) is as below:

$$\text{Min } Z = \sum_{a \in A_{pipe}} C_a \times D_a \cdot L_a \tag{6}$$

$$k_a = \beta Q_a^2 \tag{7}$$

$$k_a \geq \pi_a^j - \pi_a^i + 2y_a\left(\pi_i^{min} - \pi_j^{max}\right) \tag{8}$$

$$k_a \geq \pi_a^i - \pi_a^j + 2(y_a - 1)\left(\pi_i^{max} - \pi_j^{min}\right) \tag{9}$$

$$k_a \leq \pi_a^j - \pi_a^i + 2y_a\left(\pi_i^{max} - \pi_j^{min}\right) \tag{10}$$

$$k_a \geq \pi_a^i - \pi_a^j + 2(y_a - 1)\left(\pi_i^{min} - \pi_j^{max}\right) \tag{11}$$

$$\pi_i^{min} \leq \pi_i \leq \pi_i^{max}, \forall\ i\ \varepsilon\ N \tag{12}$$

$$\tau_a^{min} \leq \tau_a \leq \tau_a^{max}, \forall\ a\ \varepsilon\ A_{comp} \tag{13}$$

$$MQ = b \tag{14}$$

The Equations (8)–(11) are called McCormick envelopes for the equation $k_a = 2(y_a - 1)(\pi_i - \pi_j)$ [5]. These envelopes result in exact reformulation because it is the product of a binary variable with a continuous variable.

3.2. GAMS as Modeling Language

The mathematical model developed in Section 3.1 is now modeled into a computer program in General Algebraic Modeling System (GAMS) [6], which is then solved through DICOPT and CONOPT solvers. GAMS is a high-level modeling language used for mathematical optimization. DICOPT is DIscrete Continuous OPTimizer that solves Nonlinear Programming (NLP) and Mixed Integer Programming (MIP) sub-problems alternatively. CONOPT is used due to a large number of nonlinear constraints [7].

4. Computational Experiments and Results

All the computational experiments were performed on a small gas transmission network with 16 nodes and 15 arcs. The details of some arcs are shown in Table 1. Two benchmarks were created from the base network; (i) the enhancement in diameter of the segment s1 which consists of arcs 1 to 3 as well as the higher pressure at the supply nodes (ii) the installation of a compressor station at the transient node to boost the pressure. The results show that the first benchmark is feasible, whereas, the second benchmark was infeasible despite increasing the pressure. The optimal solution for the first benchmark was achieved at a 12-inch diameter [8,9].

Table 1. Arcs details.

Arc	Length (Miles)	Arc Resistance
1.2	2.78	4.45×10^{-8}
2.3	13.51	2.16×10^{-7}
3.4	2.7	4.32×10^{-8}

5. Conclusions

In this study, we developed a mathematical model with convex relaxation for the reinforcement of a gas transmission network with a minimal investment cost objective. The computational experiments showed that the developed model is an efficient tool to solve the problem of the optimal expansion of a gas transmission network in a very short time.

Author Contributions: Conceptualization, S.U. and M.N.; methodology, S.U.; software, M.N.; validation, M.N.; formal analysis, M.N.; investigation, S.U.; resources, M.N.; data curation, M.N.; writing—original draft preparation, M.N.; writing—review and editing, S.U.; visualization, M.N.; supervision, S.U.; project administration, S.U. All authors have read and agreed to the published version of the manuscript.

Funding: This research received no external funding.

Institutional Review Board Statement: Not Applicable.

Informed Consent Statement: Not Applicable.

Data Availability Statement: Not Applicable.

Conflicts of Interest: The authors declare no conflict of interest.

References

1. Bickel, T.; Himmelblau, D.; Edgar, T. Optimal Design of Gas Transmission Networks. *Soc. Pet. Eng. J.* **1978**, *18*, 96–104. [CrossRef]
2. Sundar, K.; Misra, S.; Zlotnik, A.; Bent, R. Robust Gas Pipeline Network Expansion Planning to Support Power System Reliability. In Proceedings of the 2021 American Control Conference (ACC), New Orleans, LA, USA, 25–28 May 2021; pp. 620–627.
3. Üster, H.; Dilaveroğlu, Ş. Optimization for Design and Operation of Natural Gas Transmission Networks. *Appl. Energy* **2014**, *133*, 56–69. [CrossRef]
4. Elshiekh, T.; Khalil, S.; Mawgoud, H. Optimal Design and Operation of Egyptian Gas-Transmission Pipelines. *Oil Gas Facil.* **2013**, *2*, 44–48. [CrossRef]
5. McCormick, G.P. Computability of Global Solutions to Factorable Nonconvex Programs: Part I—Convex Underestimating Problems. *Math. Program.* **1976**, *10*, 147–175. [CrossRef]
6. Li, T.; Liu, P.; Li, Z. Modelling and Optimization of a Natural Gas Supply System at a Transient Stage: A Case Study of China. *BMC Energy* **2019**, *1*, 5. [CrossRef]
7. Zheng, T.; Feng, H.; Wang, B.; Zheng, J.; Liang, Y.; Ma, Y.; Keene, T.; Li, J. Operational Optimization of a Cyclic Gas Pipeline Network with Consideration of Thermal Hydraulics. *Ind. Eng. Chem. Res.* **2021**, *60*, 2501–2522. [CrossRef]
8. Mak, T.W.K.; van Hentenryck, P.; Zlotnik, A.; Bent, R. Dynamic Compressor Optimization in Natural Gas Pipeline Systems. *INFORMS J. Comput.* **2019**, *31*, 40–65. [CrossRef]
9. dos Santos, S.P.; Eugenio Leal, J.; Oliveira, F. The Development of a Natural Gas Transportation Logistics Management System. *Energy Policy* **2011**, *39*, 4774–4784. [CrossRef]

engineering proceedings

Proceeding Paper

Driving Parameters for Technology Penetration in Pakistan: A Case Study for Indoor LED Lighting †

Ubaid Zia [1,*], Saleha Qureshi [1], Hina Aslam [1] and Muhammad Zulfiqar [2]

[1] Sustainable Development Policy Institute (SDPI), Islamabad 44000, Pakistan
[2] Department of Mechanical Engineering, Capital University of Science and Technology (CUST), Islamabad 44000, Pakistan
* Correspondence: ubaid@sdpi.org

† Presented at the 2nd International Conference on Advances in Mechanical Engineering (ICAME-22), Islamabad, Pakistan, 25 August 2022.

Abstract: This study aims to analyze the existing market status of lighting technologies in Pakistan, the existing policy and regulatory framework, and the key driving factors which have led to the market growth of LED technology in both rural and urban areas. The results obtained from an extensive survey backed by the literature on the policy landscape of energy efficiency showed that LED technology has penetrated approximately 95% of the existing market. With a low payback period and high return rates, the entire transition toward LED lighting can lower mercury pollution by 700 kg, lower carbon dioxide by 33,000 kt, and save USD 6.5 billion in the form of electricity bills. This transition has been mainly driven by the low cost of technology resulting from regulatory support in the form of the Minimum Energy Performance Standards (MEPs), labeling schemes, and reduced taxation on both sales and manufacturing.

Keywords: light-emitting diode; energy conservation; mercury pollution; technology index

Citation: Zia, U.; Qureshi, S.; Aslam, H.; Zulfiqar, M. Driving Parameters for Technology Penetration in Pakistan: A Case Study for Indoor LED Lighting. *Eng. Proc.* **2022**, *23*, 10. https://doi.org/10.3390/engproc2022023010

Academic Editors: Mahabat Khan, M. Javed Hyder, Muhammad Irfan and Manzar Masud

Published: 20 September 2022

1. Introduction

The major challenges for developing countries in their SDG pathways are the limited energy resources and resulting GHG emissions from their inefficient use [1]. Considering the rapid growth in population, urbanization, and industrialization, the role of electricity in running the whole ecosystem is extremely significant, and its irrational use could create major hurdles in achieving the targets of the sustainable development goals (SDG-7), i.e., access to reliable and affordable energy [2]. Over the past two decades, Asian countries have shown the highest growth in their energy consumption; however, at the same time, the highest share of energy wastage comes from the same countries due to their low technology development index [2].

Among the key demand sectors, lighting constitutes a significant portion of total electricity consumption in buildings. According to the Global Lighting Challenge (GCL), 15% of global electricity is consumed by the residential sector [3]. This is even greater than the total electricity produced by the entire global nuclear industry. Until 2010, the technology commonly used for lighting purposes in Pakistan was fluorescent lamps and tubes. In recent decades, most countries have made significant progress by shifting from conventional to solid-state lighting through the use of LEDs [4]. Statistics have indicated that if all existing lights were instantaneously converted to LEDs, it could save over 800 Mt of emissions globally. Another study on ASEAN countries indicated that a transition toward LEDs would result in cost savings of USD 3500 million per year while saving 35 TWh of electricity and 20 Mt of CO_2 emissions [5].

Over the product life cycle, the economic prospects of increased efficiency have been directly linked to its higher efficiency, better performance, longer lifespan, low emissions, and most importantly, cost savings in the form of reduced electricity bills [3]. Research has indicated that by 2030, Pakistan can save around 1.2 TWh by promoting energy efficiency in the lighting sector. This could further result in economic savings of around USD 120 million [2]. NEECA (MoE, Power Division) in Pakistan conducted research in collaboration with United for Energy (U4E), which indicated that a complete transition from conventional to LED lighting could save 4 TWh of annual electricity and reduce consumer bills by around USD 408 million [6].

Although the penetration of LEDs provides a more positive energy and economic outlook, it also helps achieve the targets of both energy and climate commitments. This includes Pakistan's recent support of the African Lighting Amendment (ALA), which was recently approved at the Conference of Parties of Minamata Convention on Mercury [7]. As per the convention, all parties that are signatories to the convention must phase out mercury-containing lighting products (mainly indoor) by 2024–2025. Further energy efficiency improvements aid Pakistan in achieving the targets of SDG-7 and the resulting reduced environmental emissions provide a way forward for fulfilling the commitments mentioned under Pakistan's Nationally Determined Contributions (NDCs) [8]. Therefore, this study critically analyzes the status quo of the lighting sector in Pakistan while analyzing the policy and regulatory support that was provided to LEDs, leading to its high-rate penetration.

2. Methodology and Data Collection

The methodological approach in this section consists of two data categories. Primary data were collected through a market survey of 135 shops at seven different locations in Pakistan (Islamabad, Rawalpindi, Taxila, Wah Cantt, Gujranwala, Hassanabdal, and Lahore). This survey data was used to collect the most insightful data on the lighting products available in the different categories. Table 1 indicates the sample data collection.

Table 1. Parameters for data collection (single entry mentioned here for LED bulbs).

Data Collection Parameters	
Retailer: Liaqat Electronics	Category: LED Bulb
Price: PKR 200 for 18 W LED	Brand Name: Areebah
Model: IKF D-30	Voltage and Frequency: 85–265 V * 50 Hz
Base Type: E-27	Lumens and CKI: 3000 and 6500 K
CRI and Lifetime: N/M and 8000 h	UPC: 16787300

The data collection based on the parameters defined in Table 1 was initially analyzed for identifying the market penetration of the different categories, and then it was arithmetically analyzed to calculate the energy savings, environmental savings, mercury reduction, and the payback period. Secondary data were collected through existing energy sector policies, power policies, climate change policies, and the recently published Draft National Energy Efficiency and Conservation Policy of Pakistan 2022. Furthermore, regulatory reforms, such as the Introduction of Minimum Energy Performance Standards along with the labeling scheme, were also critically analyzed and a comparison of lighting and other similar household products was also made.

3. Results and Discussion

Survey Results and Data Insights

The market analysis indicated that the low cost of LED technology has driven a rapid transition toward LED lighting in Pakistan. Figure 1a indicates the frequency with which the different categories of products were available in the market, whereas Figure 1b indicates the number of models that currently exist in the market.

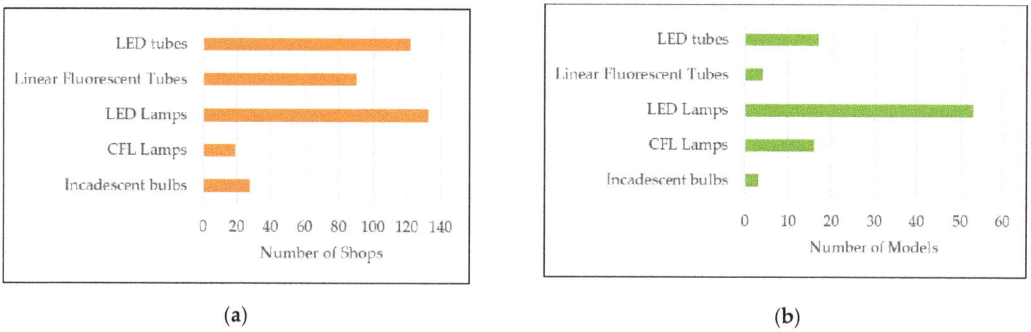

Figure 1. (**a**) Availability of different models in retail stores (**b**) Numbers of different models available.

Based on Figure 1, LEDs have the highest share in both market presence as well as the number of models present within the market. Driven by the economic analysis of the survey data, LEDs are the most feasible choice for customers.

Based on Table 2 and further analysis of the data, LED bulbs have an instantaneous payback, whereas linear tubes have a payback period of 5 months. This means that an additional investment of PKR 440 on a single product can save PKR 6900 over the life cycle of a product. Furthermore, considering the fixed quantity of light generated from different sources, LEDs can increase the rated lifetime by 12,000 h, reduce annual electricity consumption from a single bulb by 60 kWh, and reduce CO_2 emissions by around 187 kg. Through a complete transition, Pakistan can reduce mercury pollution by 700 kg, lower carbon dioxide by 33,000 kt, and reduce electricity bills by almost USD 6.5 billion. Compared to other Asian countries (India, Bangladesh, Philippines, Sri Lanka, and Vietnam), Pakistan has the lowest payback period and highest power-saving potential. These results very clearly highlight that LEDs are by far the most suitable technology for the lighting sector in Pakistan.

Table 2. Economic incentives and paybacks for the transition toward linear LEDs.

Parameter	LFTs	LED Retrofits
Lifetime	13,000 h	25,000 h
Tube Price	PKR 210/-	PKR 650/-
Power Consumption	36 Watts	18 Watts
Usage (9 h/day)	118 kWh/year	59 kWh/year
Cost of Electricity	PKR 2047/year	PKR 1037/year
Cost (7-year timeline)	PKR 14887	PKR 7192
Additional Cost payback	-	5-Months

The statistics highlighted in the results section indicate that LEDs have deeply penetrated the lighting market owing to the efforts of NEECA [9] and FBR. Customs duty for LEDs is only 3% as opposed to 20% for all other categories. Similarly, to ensure local production, the sales tax on LED products is zero, whereas it is 17% for other imported products. These interventions have led to the availability of a strong indigenous market for all available LED products. Even the larger number of LED models encountered during the survey were local products available at a much cheaper price than imported products. This has significantly driven the cost of lighting products to well below their fluorescent counterparts.

4. Conclusions

This study highlights the current market status and key drivers that have led to the country-wide adoption of LED lighting technology in Pakistan. Through a survey conducted in seven different areas in Pakistan, this study identified that LED bulbs have a sale share of 98.5% in Pakistan's retail market. CFL lamps (known previously as energy savers) despite being introduced after ICs have a lower share of sales owing to their higher cost (20.7% for ICs and 14% for CFLs). Florescent linear tubes are however still available in approximately 90% of the relevant market due to the high cost of the LED linear tubes that come with a structure intact. However, the payback analysis identified that although LED bulbs have an instantaneous payback, LED linear tubes have a payback of 5 months. Thus, spending an additional PKR 440 today can save approximately PKR 6900 over the product lifecycle. Through a complete transition from fluorescent lighting to LEDs, Pakistan can cumulatively reduce mercury pollution by 700 kg, lower carbon dioxide by 33,000 kt, and reduce electricity bills by almost USD 6.5 billion. On a regulatory and policy front, this study identified the key roles played by FBR and NEECA in reducing taxation and producing MEPs and labeling schemes for a rapid transition.

Author Contributions: Conceptualization, U.Z. and S.Q.; methodology, U.Z. and H.A.; software, U.Z. and S.Q.; validation, U.Z. and S.Q.; formal analysis, U.Z. and M.Z.; investigation, U.Z. and H.A.; data curation, M.Z.; writing—original draft preparation, U.Z.; writing—review and editing, U.Z. and H.A.; supervision, H.A.; project administration, U.Z. All authors have read and agreed to the published version of the manuscript.

Funding: This research received no external funding.

Institutional Review Board Statement: Not applicable.

Informed Consent Statement: Not applicable.

Data Availability Statement: Please refer to data available at: https://bit.ly/3aHQ0tm (accessed on: 21 March 2021).

Conflicts of Interest: The authors declare no conflict of interest.

References

1. Galvin, R. Net-Zero-Energy Buildings or Zero-Carbon Energy Systems? How Best to Decarbonize Germany's Thermally Inefficient 1950s-1970s-Era Apartments. *J. Build. Eng.* **2022**, *54*, 104671. [CrossRef]
2. Islam, N.U.; Usman, M.; Jamil, T. Energy-Savings Using Solid-State Lighting: A Case Study of India, Pakistan, and Bangladesh. *Energy Policy* **2022**, *160*, 112676. [CrossRef]
3. Aslam, H.; Ur, U.; Zia, R.; Qureshi, S. *Market Transition to LED Lighting: A Case for Energy Efficiency and Conservation In Pakistan*; Sustainable Development Policy Institute: Islamabad, Pakistan, 2022.
4. Maxson, P.; Bender, M.; Culver, A. Mercury in Fluorescent Lighting. Clean Light. Coalit. 2022. Available online: https://www.clasp.ngo/wp-content/uploads/2021/10/Mercury-in-Fluorescent-Lighting_FINAL.pdf (accessed on 2 February 2022).
5. UNEP. *ASEAN Commits to Switch to Efficient Lighting*; UNEP: Nairobi, Kenya, 2015.
6. United Nations Environment Programme. Accelerating the global adoption of energy-efficient lighting. *United Nations Environ. Program.* **2017**, *132*, 409–423. Available online: http://hdl.handle.net/20.500.11822/20406 (accessed on 10 September 2021).
7. CLiC. Convention on Mercury Agrees to Phase out Major Category of Fluorescent Light Bulbs, but Last-Minute Interventions Delay Action on Another. 2022. Available online: https://cleanlightingcoalition.org/news/convention-on-mercury-agrees-to-phase-out-major-category-of-fluorescent-light-bulbs-but-last-minute-interventions-delay-action-on-another/ (accessed on 10 May 2022).
8. MoCC. *Pakistan Updated NDC 2021*; Government of Pakistan: Islamabad, Pakistan, 2021.
9. NEECA. *Launching of MEPs and Energy Labeling Scheme*; NEECA: Islamabad, Pakistan, 2020.

engineering proceedings

MDPI

Proceeding Paper

Experimental Investigation of Runner Design Parameters on the Performance of Vortex Turbine †

Rizwan Ullah * and Taqi Ahmad Cheema *

Faculty of Mechanical Engineering, GIK Institute of Engineering Sciences and Technology, Topi 23640, Pakistan
* Correspondence: rizwanullah@giki.edu.pk (R.U.); tacheema@giki.edu.pk (T.A.C.)
† Presented at the 2nd International Conference on Advances in Mechanical Engineering (ICAME-22), Islamabad, Pakistan, 25 August 2022.

Abstract: The vortex turbine (VT) is a micro-hydro turbine that extracts power from a water vortex that is artificially generated in a conical or cylindrical cross-section basin. The former cross-section gives rise to a stronger vortex than the latter, meaning it has more potential for power generation. For this reason, the present study experimentally analyzed the effect of sensitive geometric parameters on the VT performance, i.e., the rotor to basin diameter ratio (RB) and runner's position in the conical basin (CB). The results show that the ideal runner in terms of RB for the best performance of VT is a runner with RB = 0.6, which has a maximum potential for the utilization of the forced vortex region of the Rankine vortex. Moreover, the best position for the installation of a VT runner is the location at the bottom near the orifice, as the strong vortex tangential velocity and maximum head drop at the mid-position is not a feasible option. Blades with a tilt in the vertical plane are suggested for use in the power extraction at the bottom of CB whereas crossflow blades suit the rotational flow region near the top of CB, i.e., the surface vortex.

Keywords: gravitational water vortex turbine; potential flow; runner position; conical basin; Rankine vortex

Citation: Ullah, R.; Cheema, T.A. Experimental Investigation of Runner Design Parameters on the Performance of Vortex Turbine. *Eng. Proc.* **2022**, *23*, 14. https://doi.org/10.3390/engproc2022023014

Academic Editors: Mahabat Khan, M. Javed Hyder, Muhammad Irfan and Manzar Masud

Published: 20 September 2022

1. Introductions

A huge amount of hydro energy has remained unutilized due to low flowrate and a low-head issue [1]. This problem can be addressed with the use of gravitational water vortex turbine (VT) technology, as all other conventional hydro turbines need one of the above-mentioned parameters in a sufficient amount. This type of turbine can easily operate from 0.7 m to 3 m under the water head [2]. It is used to extract energy from the water vortex, which is artificially generated when the water exits under the force of gravity from the orifice of the basin in either a cylindrical or conical configuration [3]. The VT performance reduces when the blades are varied from 6 to 12 on a runner [4]. The performance of the VT has also been analyzed with different profile blade runners, such as centrifugal, Francis, and impulse paddle type, while it is suggested that the runner should have a minimum hub diameter with straight blades at an angle for better VT performance [3,5]. It is reported that the CB provides a stronger vortex than a cylindrical basin under the same water head and flow rate [4]. Moreover, the vortex can re-originate itself in the CB once it is distorted through a runner that is fixed in the basin. Therefore, for the first time, Gheorghe-Marius et al. proposed the idea of using stepwise runners in a CB, and they reported a detailed analytical model [3]. Later, the same approach was used for the purpose of a detailed experimental investigation of multi-stage VT performance [6]; The authors' recent study is particularly focused on the inter-stage and intra-stage performance evaluation of multi-stage VT [7].

This study is especially devoted to the sensitive runner's design parameters, including the rotor diameter to corresponding basin diameter ratio (RB) and runner's position in the CB for the purpose of maximum power extraction. The effect of the above parameters on the performance of the VT has been examined by plotting the brake Torque (T), power (P), and efficiency (η) vs. the rotational speed (rpm). To the best of our knowledge, the present manuscript is of worthy consideration for the further exploration of VT technology.

2. Experimental Setup and Methodology

The designed experimental rig used for the performance evaluation of the VT is shown in Figure 1a. The testing facility comprised a centrifugal pump, storage tank, overhead reservoir, conical basin, and assembly of a VT. The VT runner, which had four blades of the Savonius profile, was used to extract power from the water vortex. The bucket method was used for the measurement of flowrate [7]. An inlet flow rate of 4 L/s was kept constant for all the experiments. A tachometer and Prony brake mechanism were used for the measurement of rotational speed, N (rpm), and brake torque, T (N-m), respectively.

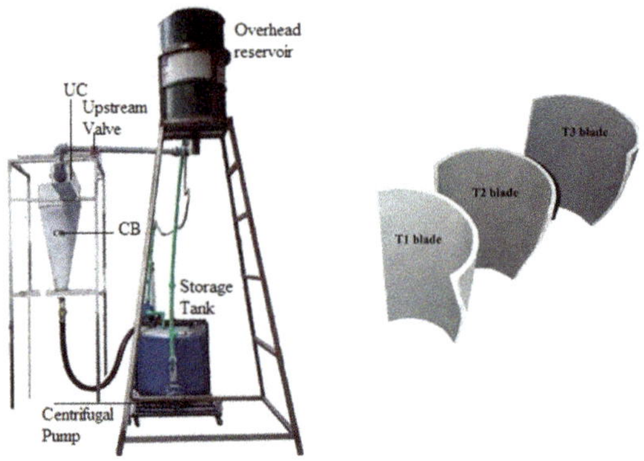

(**a**) (**b**)

Figure 1. (**a**) Experimental test rig for the VT PPs investigation; (**b**) the Savonius blades profile.

The following equations are used for the measurement of the experimental angular velocity ω (rad/s), brake power P (watt), and turbine efficiency η.

$$\Omega = 2\pi N/60 \quad (1)$$

$$P = T\omega \quad (2)$$

$$\eta = P/HP \quad (3)$$

"HP" refers to the available hydraulic power. The developed mathematical model is used for the analytical measurements of the above performance parameters (PPs) [7].

3. Results and Discussion

In the operation of the turbine, it is important to know the value of torque and RPM for a specific power output under a constant water head and inlet flow rate. Therefore, the effect of the RB and position of a runner on the performance of the VT have been depicted by plotting the performance curves at various load conditions.

3.1. Analysis for Rotor Diameter to Corresponding Basin Diameter Ratio (RB)

The RB is among the major geometric parameters that greatly influence the PPs of the VT. The variation in the RB results in the change of the runner's diameter. For this reason, the RB is varied by the fixing of the blades of Figure 1b, which were T1, T2, and T3 at stage 3 (S3), as shown in Figure 2. The first subscript terms R31, R32, and R33 stand for the stage of the VT, while the second subscript stands for the type of blade. For example, R31 is the runner fixed at stage 3 while having T1 blades. In Figure 2, the comparison between R33 and R31 showed that R33 performed better than R31 with a higher brake torque (38%↑)

at the cost of low rpm (17%↓), thus providing more power. Under the same conditions, R32 competes with R33 in the same power production and efficiency with an advantage of less weight; thus, it may be termed as the optimum performing runner when RB = 0.6. Figure 2 also reflects that the VT runner with RB = 0.47, showing its installation within the rotational flow region of the Rankine vortex; it achieved the highest rpm. In the Rankine vortex, the rotational vortex has a greater ability for power generation due to the existence of a strong tangential velocity in it, unlike the irrotational flow region [8]. The maximum RB runner reflects its presence in the region of both free and forced vortex while providing the lowest rpm.

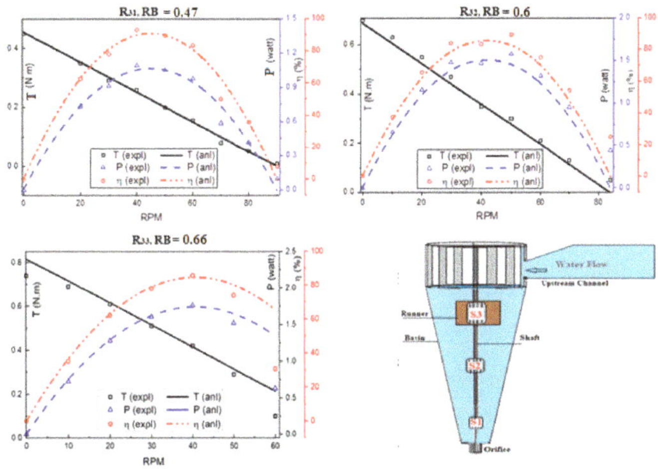

Figure 2. PPs of the VT with different RB runners and schematic of VT setup.

The optimum performer runner (R32) implies its installation on the interface of the rotational and irrotational vortex. R32, in terms of PPs, overshoots the analytical calculations due to having the highest capacity for the utilization of the rotational vortex region. Thus, based on the discussion above, a runner with RB = 0.6 is suggested as a runner for VTs.

3.2. *Effect of Runner Position*

The mounting position of an individual runner in a CB has a great impact on the performance of a VT. At the same input design parameters, the maximum energy can be harnessed from the vortex by fixing the turbine runner at its proper location along the height of the shaft. This is why the position of the runner is varied by fixing the T1 blades of Figure 1 at S1, S2, and S3, respectively (Figure 2). Thus, the insight runners in this section are R11, R21, and R31.The results achieved through the adopted technique are plotted in Figure 3.

The highest rotational speed of R11 at S1 is because of the maximum head drop at the same inlet flowrate. Among all the runners, the runner fixed at S2 (R21) underperformed the other two, R11 and R31, in all modes of PPs. Thus, it can be suggested that the mid-position of CB should not be utilized for power generation in the case of single-stage turbines. Between the runners R11 and R31, the more promising candidate is R11. It produced double the power compared with R31 with a lower water head utilization tendency, as indicated by its lower efficiency. On the other hand, R31 has more competency for water head utilization over R11, thus making it difficult to credit the S1 or S3 position for energy generation. An in-depth observation would shed an inkling on the selection of the blades profile. S1 and S3 are the location of the maximum and minimum head drops, respectively. Thus, we suggest that blades of the same profile are tilted in a vertical plane at S1 to utilize both head and vortex tangential velocity. The vortex tangential velocity resulting through the water head drop and artificial vortex generation in the CB is the major contributor to the higher power of R11, which is fixed near the bottom. In the position of S3, the available water head is minimum, but the blades of the selected profile suit S3 well. We conclude that in the case of a single-stage VT, the bottom position near the orifice while using the runner with blades tilted in a vertical plane is preferred for power generation. The top position is feasible for power generation if the runners have blades of the present profile.

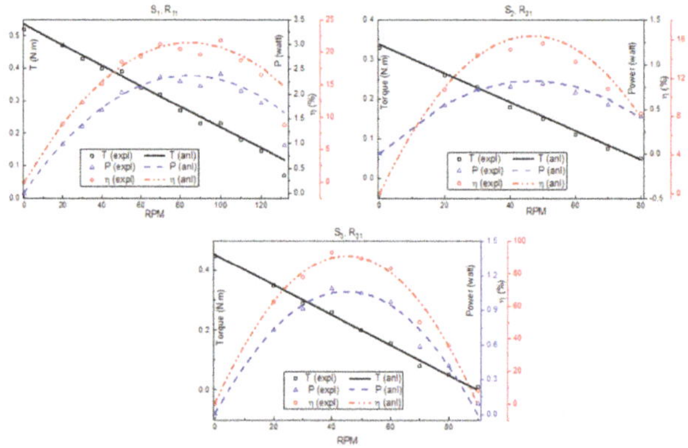

Figure 3. PPs of the VT runners with T1 blades positioned at S1, S2, and S3, respectively.

4. Conclusions

The purpose of the present study is to trace the best diameter (RB) runner and fixing position of the vortex turbine in a conical cross-section basin for maximum power output. The analysis was carried out based on rotational speed (rpm), power (watt), and efficiency under different loads. For the best performance of the VT, it is recommended to use a runner with RB = 0.6 as it showed the maximum potential in utilizing the potential flow in the Rankine vortex. The bottom position near the orifice is preferred for the installation of the VT due to the presence of a strong vortex tangential velocity.

Author Contributions: Conceptualization, methodology, analysis, and writing of this work is carried out by R.U. while reviewing, editing, and supervision has been done by T.A.C. All authors have read and agreed to the published version of the manuscript.

Funding: This research received no external funding.

Institutional Review Board Statement: Not Applicable.

Informed Consent Statement: Not Applicable.

Data Availability Statement: Not applicable.

Acknowledgments: The authors acknowledge the technical and financial support of the GIK Institute of Engineering Sciences and Technology, Topi, Pakistan.

Conflicts of Interest: The authors declare no conflict of interest.

References

1. Date, A.; Akbarzadeh, A. Design and Cost Analysis of Low Head Simple Reaction Hydro Turbine for Remote Area Power Supply. *Renew. Energy* **2009**, *34*, 409–415. [CrossRef]
2. Zotloterer: Smart-Energy-System. Gravitation Water Vortex Power Plant. Available online: http://www.zotloeterer.com/welcome/gravitation-water-vortex-power-plants/ (accessed on 7 July 2022).
3. Gheorghe-Marius, M.; Tudor, S. Energy Capture in the Gravitational Vortex Water Flow. *J. Mar. Technol. Environ.* **2013**, *1*, 89–96.
4. Dhakal, S.; Thapa, A.; Nakarmi, S.; Pun, P.; Bajracharya, T. Development and Testing of Runner and Conical Basin for Gravitational Water Vortex Power Plant. *J. Inst. Eng.* **2014**, *10*, 140–148. [CrossRef]

5. Saleem, A.S.; Ullah, R.; Cheema, T. Experimental Investigation of Various Blade Configurations of Gravitational Water Vortex Turbine (GWVT). In Proceedings of the International Conference on Power Generation Systems and Renewable Energy Technologies (PGSRET), Islamabad, Pakistan, 10–12 September 2018.
6. Ullah, R.; Cheema, T.A.; Saleem, A.S.; Ahmad, S.M.; Chattha, J.A.; Park, C.W. Performance Analysis of Multi-Stage Gravitational Water Vortex Turbine. *Energy Convers. Manag.* **2019**, *198*, 111788. [CrossRef]
7. Ullah, R.; Cheema, T.A.; Saleem, A.S.; Ahmad, S.M.; Chattha, J.A.; Park, C.W. Preliminary Experimental Study on Multi-Stage Gravitational Water Vortex Turbine in a Conical Basin. *Renew. Energy* **2020**, *145*, 2516–2529. [CrossRef]
8. Giaiotti, D.B. The Rankine Vortex Model: PhD Course on Environmental Fluid Mechanics. Ph.D. Thesis, University of Trieste, Trieste, Italy, 2006.

engineering proceedings

MDPI

Proceeding Paper

Design and Airflow Analysis of 20 kW Horizontal Axis Wind Turbine Blade †

Muhammad Hamza [1], Ahmad Ali [1], Saqlain Abbas [1,*] and Zulkarnain Abbas [2]

[1] Department of Mechanical Engineering, University of Engineering and Technology Lahore (Narowal Campus), Narowal 51750, Pakistan
[2] Department of Mechanical Engineering, National Fertilizer Cooperation (NFC), Institute of Engineering and Technology (IET), Multan 60000, Pakistan
* Correspondence: Saqlain.abbas@uet.edu.pk
† Presented at the 2nd International Conference on Advances in Mechanical Engineering (ICAME-22), Islamabad, Pakistan, 25 August 2022.

Abstract: The key objective of the research is to calculate and design the Small Horizontal Axis Wind Turbine (HAWT) that can meet Pakistan's energy needs. This is the plan for producing approximately 20 kW of electricity to distribute the load used by common household appliances. This study will focus on Jamshoro, Sindh, Pakistan, where a wind turbine is considered to generate electricity. The appropriate design is required to make the turbine more efficient and decreases the cost. Q-blade wind turbine software verifies the design parameters. The maximum power factor is achieved at the design speed of 8 m/s. Design analysis is also performed in Q-blade wind turbine simulation software.

Keywords: horizontal axis wind turbines; power factor; Q-blade software

Citation: Hamza, M.; Ali, A.; Abbas, S.; Abbas, Z. Design and Airflow Analysis of 20 kW Horizontal Axis Wind Turbine Blade. *Eng. Proc.* **2022**, *23*, 11. https://doi.org/10.3390/engproc2022023011

Academic Editors: Mahabat Khan, M. Javed Hyder, Muhammad Irfan and Manzar Masud

Published: 20 September 2022

1. Introduction

Pakistan is among the ten countries in the world facing the most serious energy crisis. As the country faces challenges with numerous monetary, political, and social issues, the conversion from a traditional fill-based economy to a green economy will be difficult. The household consumes 30% of the total energy generated and electricity demand is increasing daily and posing a challenge to the current energy solution. In urban areas, the energy demand increases more as people migrate from the big city [1].

In the old days, the energy produced from the wind was used for many purposes as the grinding of wheat and corn. However, "Blade Element Momentum Theory" was introduced during the 1980s. Using this theory, many wind turbine designs were presented [2]. The energy demand in the world is increasing continuously, and 2% of this demand is increasing through fossil fuel, which has terrible and adverse effects on our environment [3]. Until 2009, not a single network of wind farms existed. However, presently, the circumstance has changed remarkably, and wind ranches adding to the general framework are now a reality [4,5]. Presently, the current administration of Pakistan has introduced an arrangement for a power age up to 2030 that assumes an introduced capacity in the north of 162 GW [6]. In the south (Sindh and Baluchistan), Pakistan has a 1046 km coastline, although most wind power projects are now being built in the GharoKeti Bander and Hyderabad wind corridors. Calculating the axial and circumferential inducible factors of wind turbine wings is challenging. The geometry of the blades is very complex; additionally, the blades are now manufactured with glass fiber and carbon fiber composite layer structures for high wind turbine performance [7,8].

The purpose of the research is to calculate and design a small wind turbine for small-scale energy needs in Pakistan. This is the plan for generating about 20 kW of electrical energy to spread the load used by common household machines. The area for this study is

Jamshoro, Sindh province, Pakistan, the place where wind turbines will be built to generate power. Wind power design formulations are used to estimate the blade design values. At a designed velocity of 8 m/s, a reliable energy coefficient is obtained.

2. Design Parameters

The following parameters mentioned in Table 1 are analyzed and considered for the blade design.

Table 1. Design Parameters.

Blade Length (l)	Tower Height (h)	Swept Area (A)	Rotor Diameter (d)	Rotor Speed (ω)	Number of Blades (n)
7.2 m	40 m	162.86 m^2	6.92 m	89.66 rpm	3

3. Results and Discussion

The direction of the wind is one of the wind aspects. The location of a wind farm and the placement of wind turbines inside the wind farm are heavily reliant on statistical information about wind directions over time. Upwind wind turbines reduce wind speed and increase disturbance for downwind wind turbines. The wind's power is directly proportional to the cubic of mean wind speed (v).

$$\mathrm{P} \propto \mathrm{v}^3 \tag{1}$$

Output Power is the electric power supplied by the generator [9].

$$\mathrm{P} = \frac{1}{2}\mathrm{C_P}\rho\mathrm{A}\mathrm{v}^3 \tag{2}$$

The value of $\mathrm{C_P}$ (Power Coefficient) decreases as wind speed increases. Equation (3) shows the relation between $\mathrm{C_P}$ and wind speed.

$$\mathrm{C_P} \propto \frac{1}{\mathrm{v}^3} \tag{3}$$

The tangential speed at the tip of the blade is divided by the actual wind speed to calculate TSR. Its value determines by using Equation (4). Tip speed ratios concerning different wind speeds relation is shown in Table 2 at different wind speeds.

$$\lambda = \frac{\omega \mathrm{r}}{\mathrm{v}} \tag{4}$$

Table 2. Output power.

Wind Speed (m/s)	$\mathrm{C_P}$	Tip Speed Ratios (λ)	Output Power (kW)
7	0.58	9.2	19.9
7.5	0.47	8.67	19.84
8	0.42	8.12	21.538

Actual output can be determined by Equation (2).

4. Airfoil

Airfoils for wind turbines (HAWTs) are often designed with high-lift coefficients and low-drag coefficients for use at small angles of attack. S1020 airfoil | Ornithopter airfoil is used in this research because it gives a reliable lift-to-drag ratio which will be useful to stabilize the wind turbine. Figure 1 shows the shape of the S1020 airfoil.

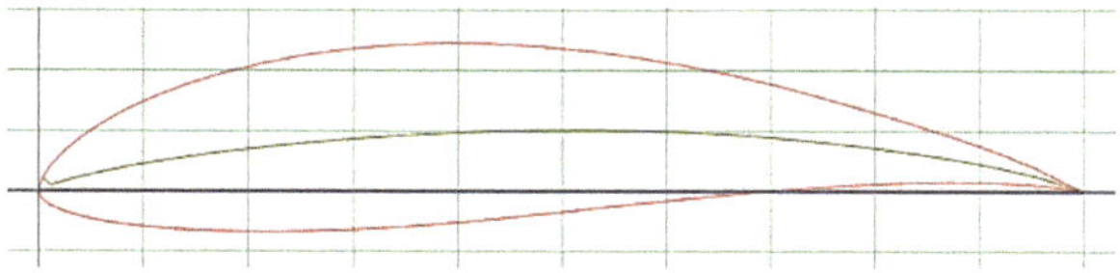

Figure 1. Airfoil design.

4.1. Foil Specifications

As it increases, the value of the Alpha (Angle of Attack) ratio of lift coefficient to drag coefficient starts decreasing. In real-life, Alpha is between 10 to 15 degrees [10]. In this research, values of Lift to Drag ratio are analyzed for different Angles of Attack between 0° to 12° degrees; it gives very smooth values to 12 degrees. Figure 2a,b show the specification of the airfoil S1020 at 12 degrees and the graph between Alpha and Cl/Cd.

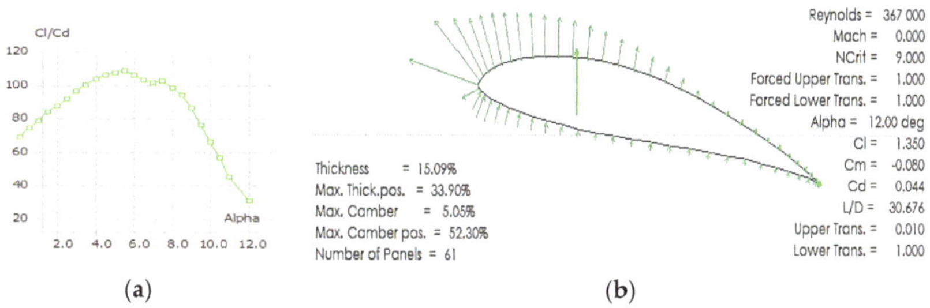

Figure 2. Alpha (Angle of Attack) and ratio of lift coefficient to drag coefficient.

4.2. Static Structural Analysis

Static Structural Analysis analyzes the effect of stable (or stationary) loading states on a body, while inertia and the damping effects induced by time variable loads are neglected. In this research, the Wind turbine blade is fixed at a circular foil end which is a boundary condition. Figure 3 simulates the Static Structural Analysis for the turbine.

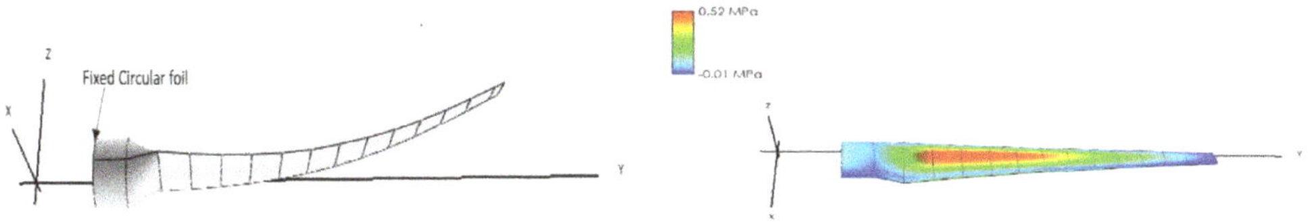

Figure 3. Static Structural Analysis.

4.3. Airflow Analysis

Airflow analysis is conducted using Q-blade software and analyzed at 8 m/s since using equation (2) approximately 20 kW of electricity can be achieved, theoretically. The colors on the blades show stress on the blades during airflow. By using software, calculated values can be observed experimentally. The calculations are practically testified by using Q-blade wind turbine software at 8 m/s, HWAT generates more than 20 kW of electricity. The airflow analysis and output calculations as described in Figure 4.

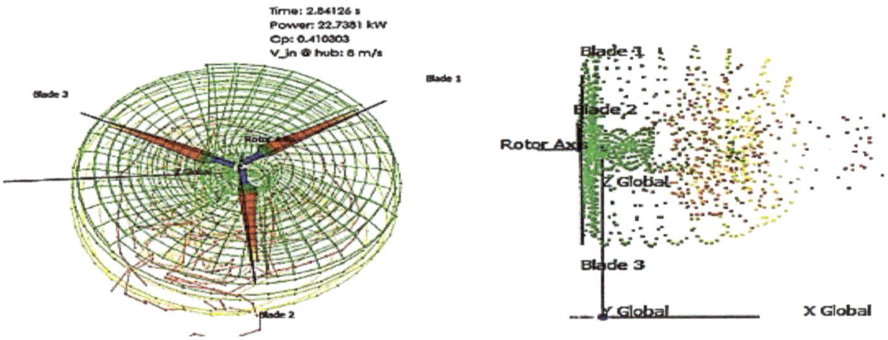

Figure 4. Airflow Analysis of the wind turbine blade.

5. Conclusions

This study demonstrates the precise design and calculation of 20 kW horizontal axis wind turbines to meet Pakistan's energy demands—on a small scale. The power factor can be deduced from the current design and calculation. It varies slightly according to wind speed, and its value rises at low wind speeds compared to high wind speeds. In addition, airflow analysis is conducted using Q-blade software at 8 m/s. Hence, the proposed design is suitable to provide pollution-free and sustainable energy.

Author Contributions: Conceptualization, M.H. and S.A.; methodology, M.H.; software A.A.; validation, S.A., A.A. and Z.A.; formal analysis, Z.A. All authors have read and agreed to the published version of the manuscript.

Funding: This research received no external funding.

Informed Consent Statement: Not applicable.

Data Availability Statement: Not applicable.

Acknowledgments: The authors are thankful to UET Lahore, Pakistan for providing the opportunity to finish this research work.

Conflicts of Interest: The authors declare no conflict of interest.

References

1. Chougule, P.; Nielsen, S. Overview and Design of Self-Acting Pitch Control Mechanism for Vertical Axis Wind Turbine Using Multi-Body Simulation Approach. *J. Phys. Conf. Ser.* **2014**, *524*, 012055. [CrossRef]
2. Khan, S.; Shah, K.; Izhar-Ul-Haq; Khan, H.; Ali, S.; Ahmad, N.; Abid, M.; Ali, H.; Ihsanullah; Sher, M. Observation of the Starting and Low Speed Behavior of Small Horizontal Axis Wind Turbine. *J. Wind Energy* **2014**, *2014*, 527198. [CrossRef]
3. Porté-Agel, F.; Bastankhah, M.; Shamsoddin, S. Wind-Turbine and Wind-Farm Flows: A Review. *Bound. Layer Meteorol.* **2020**, *174*, 1–59. [CrossRef]
4. Siddique, S.; Wazir, R. A Review of the Wind Power Developments in Pakistan. *Renew. Sustain. Energy Rev.* **2016**, *57*, 351–361. [CrossRef]
5. Ghafoor, A.; ur Rehman, T.; Munir, A.; Ahmad, M.; Iqbal, M. Current Status and Overview of Renewable Energy Potential in Pakistan for Continuous Energy Sustainability. *Renew. Sustain. Energy Rev.* **2016**, *60*, 1332–1342. [CrossRef]
6. Abbas, Z.; Abbas, S.; Butt, Z.; Pasha, R.A. Design and Parametric Investigation of Horizontal Axis Wind Turbine. *J. Shanghai Jiaotong Univ. (Sci.)* **2018**, *23*, 345–351. [CrossRef]
7. Lanzafame, R.; Messina, M. Horizontal Axis Wind Turbine Working at Maximum Power Coefficient Continuously. *Renew. Energy* **2010**, *35*, 301–306. [CrossRef]
8. Baloch, M.H.; Kaloi, G.S.; Memon, Z.A. Current Scenario of the Wind Energy in Pakistan Challenges and Future Perspectives: A Case Study. *Energy Rep.* **2016**, *2*, 201–210. [CrossRef]

9. Sohoni, V.; Gupta, S.C.; Nema, R.K. A Critical Review on Wind Turbine Power Curve Modelling Techniques and Their Applications in Wind Based Energy Systems. *J. Energy* **2016**, *2016*, 8519785. [CrossRef]
10. Osei, E.Y.; Opoku, R.; Sunnu, A.K.; Adaramola, M.S. Development of High-Performance Airfoils for Application in Small Wind Turbine Power Generation. *J. Energy* **2020**, *2020*, 9710189. [CrossRef]

engineering proceedings

MDPI

Proceeding Paper

Model Development for Discharge Data Extension for Ungauged Rivers Channels: A Case Study of the Proposed River Orle Hydropower Plant †

Luqman Muhammed Audu *, Nicholas Akhaze Musa, Abdulkarim Nasir and Muhammadu Masin Muhammadu

Mechanical Engineering Department, Federal University of Tech, Minna 920211, Nigeria
* Correspondence: luqslama@gmail.com
† Presented at the 2nd International Conference on Advances in Mechanical Engineering (ICAME-22), Islamabad, Pakistan, 25 August 2022.

Abstract: The paper presents a model to carry out a short-term flow data extension for a minimum of 30 years using the Gauss–Newton Empirical Regression Algorithm (GNRA) for the determination of the hydropower generation capacity of rivers in ungauged channels. An averaged 2 years of precipitation, observed experimental discharged data, and 30 years of historical and predictive precipitation data were used to generate a regression model equation after authentication analysis. A minimum, average, and maximum of 30 years of historical, and predictive discharge data and power characteristics of the river were generated. A discharge predictive accuracy of 96.71% and a Pearson Correlation Coefficient of 0.954 were established between the experimental and model results. The river has minimum, average, and peak power potentials of 5 MW, 10 MW, and 20 MW, respectively, and is capable of yielding power throughout the year.

Keywords: data extension; discharge; rivers; hydro; power; regression; analysis

Citation: Audu, L.M.; Musa, N.A.; Nasir, A.; Muhammadu, M.M. Model Development for Discharge Data Extension for Ungauged Rivers Channels: A Case Study of the Proposed River Orle Hydropower Plant. *Eng. Proc.* **2022**, *23*, 27. https://doi.org/10.3390/engproc2022023027

Academic Editors: Mahabat Khan, M. Javed Hyder, Muhammad Irfan and Manzar Masud

Published: 22 September 2022

1. Introduction

The need to improve Nigeria's power generation has emphasized the importance of improving the country's hydropower generation output [1]. However, the development of hydropower resources is currently being hampered by a hydrological data shortage due to large ungauged river channels. Data extension techniques using empirical rainfall-runoff models are used to overcome this challenge [2].

A recent study [3] developed a new hybrid biogeography-based optimization (BBO) technique to achieve a better capability to predict daily stream flow. The study referenced in [4] applied regression analysis and clustering in catchments to give excellent results in the data analysis and forecasting of hydrological runoff; the study referenced in [5] also used the regression tree ensemble approach to develop a model with good accuracy for runoff prediction. Ramana validated the adequacy of regression analysis for runoff prediction with the formulation of a model with three hydrological modules [6]. The present methods of modeling surface runoff involve complex evaluation processes with many interconnecting variables [7] that are not available for most river basins in Nigeria.

The paper presents a faster, cheaper, and more convenient model to carry out short-term flow data extension to a minimum of 30 years using the Gauss–Newton Regression Algorithm (GNRA) for data extension in ungauged river channels. GNRA is one of the most popular, efficient, and simple methods for solving non-linear problems [8].

2. Materials and Methods

The observed 2 years of experimental discharge data were extended to 30 years of discharge data using GNRA. The regression model equation was generated and validated using rainfall data for 2018 and 2019 for the study period. The validated model equation was used to carry out 30 years of the data extension process. Correlation analysis was implemented between the model and experimental results.

Modeling Equations

The general relationship for the empirical modeling process is shown in Equation (1).

$$Q = f[X, Y] \quad (1)$$

where

Q is the runoff output;
X is the input dataset of rainfall;
Y is the input dataset of historical runoff.

The regression model equation generated by the study process is given in Equation (2).

$$\mathrm{D} = 6.5870 + 0.2614\mathrm{F} - 0.0042\mathrm{F}^2 + 0.000019\mathrm{F}^3 \quad (2)$$

D is the discharge (m^3/s);
F is the rainfall (mm).

The hydroelectric power output was evaluated using Equation (3).

$$\mathrm{P} = \rho g h D \eta \quad (3)$$

where

P is the generated power (MW);
ρ is the density of water (1000 kg/m^3);
g is the acceleration due to gravity (9.81 m^2/s);
h is the head of water (50 m);
η is the plant efficiency (0.90).

3. Results and Discussions

3.1. River Orle Rainfall–Discharge Regression Output Analysis

The River Orle rainfall–discharge output is shown in Figure 1. It indicates a good fit, with the data point close to the regression line plot. This indicates low values of residuals, which equally indicates the agreement between the model and experimental results. Figure 2 represents the plot of the observed experimental results and the regression model output.

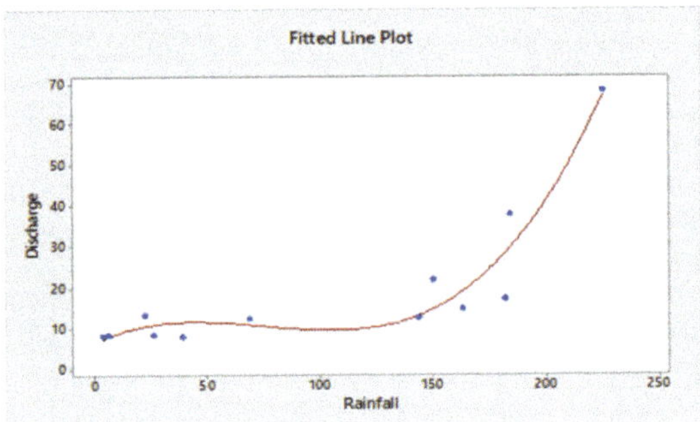

Figure 1. Fitted rainfall–discharge line plot.

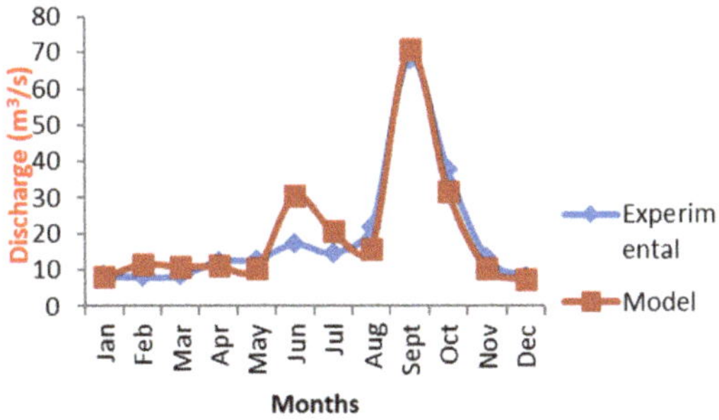

Figure 2. Profile of experimental and model discharge plot.

The mean annual discharge for the experimental and model results are 19.282 m^3/s and 19.937 m^3/s, respectively, whereas the total average discharges are 231.393 m^3/s and 239.252 m^3/s, respectively. This indicates a model discharge predictive accuracy of 96.71%. The large difference in the experiment and model discharge for June and July is because the model data were derived from 60 years of monthly rainfall data, whereas the experimental data were derived from 2 years of average monthly data. This is because there used to be occasional high-volume rainfall between June and July for a number of years.

The proximity in the discharge indicates that the monthly discharge of the river falls within the same range and demonstrates the accuracy and precision of the model.

3.2. *Power Generation Analysis*

Figure 3 indicates the discharge and power generation profile of the river. The medium flow range of the river, Q_{10} to Q_{70}, is between 10 m^3/s and 42 m^3/s which corresponds to the 5 MW–20 MW power output range. The river has a peak power output of 20 MW, base power of 10 MW, and low power of 5 MW, respectively, within this hydropower power generation range.

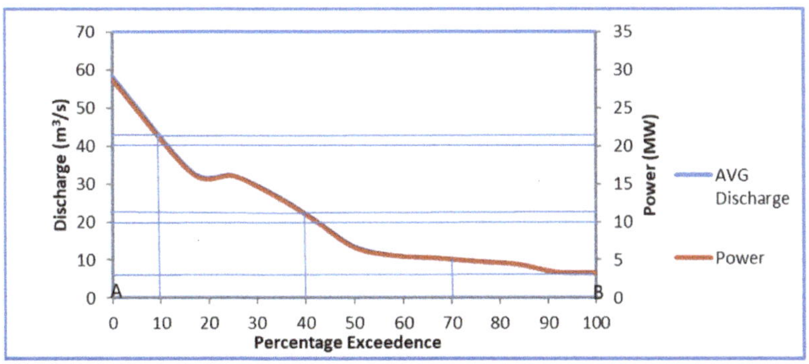

Figure 3. Integrated discharge and power duration curve.

4. Conclusions

A 30-year historical and predictive discharge data extension was carried out using the Gauss–Newton Regression Algorithm from the observed 2 years of experimental discharge data, in order to meet the requirements for the design of hydropower facilities. The profile of the river's average monthly discharge and power generation output was determined with the generated and validated model. Model discharge predictive accuracy of 96.71% and a Pearson Correlation Coefficient of 0.954 were established between the experimental and model results. The river has minimum, average, and peak power potentials of 5 MW,

10 MW, and 20 MW, respectively, within the medium flow range, and is capable of yielding power throughout the year.

Author Contributions: Conceptualization, N.A.M. and A.N.; methodology, L.M.A. and M.M.M.; software, L.M.A. and A.N.; validation, L.M.A. and A.N.; formal analysis, L.M.A.; investigation, L.M.A.; data curation, N.A.M.; writing—original draft preparation, L.M.A.; writing—review and editing, L.M.A. and M.M.M.; supervision, N.A.M. and A.N.; project administration, L.M.A. All authors have read and agreed to the published version of the manuscript.

Funding: This research received no external funding.

Institutional Review Board Statement: Not applicable.

Informed Consent Statement: Not applicable.

Data Availability Statement: Historical, predictive, and study period meteorological data were obtained from the National Centre for Meteorological Research (CNRM) [9].

Conflicts of Interest: The authors declare no conflict of interest.

References

1. Abdulkadir, T.; Salami, A.; Anwar, A.; Kareem, A. Modelling of Hydropower Reservoir Variables for Energy Generation: Neural Network Approach. *Ethiop. J. Environ. Stud. Manag.* **2013**, *6*, 310–316. [CrossRef]
2. Pechlivanidis, I.G.; Jackson, B.M.; Mcintyre, N.R.; Wheater, H.S. Catchment Scale Hydrological Modelling: A Review of Model Types, Calibration Approaches and Uncertainty Analysis Methods in the Context of Recent Developments in Technology and Applications. *Glob. NEST J.* **2011**, *13*, 193–214.
3. Roy, B.; Singh, M.P. An Empirical-Based Rainfall-Runoff Modelling Using Optimization Technique. *Int. J. River Basin Manag.* **2020**, *18*, 49–67. [CrossRef]
4. Mishra, S.; Saravanan, C.; Dwivedi, V.K.; Shukla, J.P. Rainfall-Runoff Modeling Using Clustering and Regression Analysis for the River Brahmaputra Basin. *J. Geol. Soc. India* **2018**, *92*, 305–312. [CrossRef]
5. Zhang, Y.; Chiew, F.H.S.; Li, M.; Post, D. Predicting Runoff Signatures Using Regression and Hydrological Modeling Approaches. *Water Resour. Res.* **2018**, *54*, 7859–7878. [CrossRef]
6. Ramana, G.V. Regression Analysis of Rainfall and Runoff Process of a Typical Watershed. *Int. J. Sci. Appl. Inf. Technol.* **2014**, *3*, 16–26.
7. Sitterson, J.; Knightes, C.D.; Parmar, R.; Wolfe, K.; Avant, B.; Muche, M.E. An Overview of Rainfall-Runoff Model Types. In Proceedings of the 9th International Congress on Environmental Modelling and Software, Fort Collins, CO, USA, 24–28 June 2018.
8. Siregar, R.W.; Tulus; Ramli, M. Analysis Local Convergence of Gauss-Newton Method. In Proceedings of the IOP Conference Series: Materials Science and Engineering, Medan, Indonesia, 21–23 August 2017; Institute of Physics Publishing: Bristol, UK, 2018; Volume 300. [CrossRef]
9. Centre National de Recherches Météorologiques Softwares and Open Data—National Centre for Meteorological Research. Available online: https://www.umr-cnrm.fr/spip.php?rubrique76&lang=en (accessed on 30 August 2022).

engineering proceedings

Proceeding Paper

Monthly Average Solar Radiation and Angstrom-Prescott Model for Islamabad, Pakistan †

M. Javed Hyder, Asadullah Nauman * and Shams Musheer

Department of Mechanical Engineering, Capital University of Science and Technology (CUST), Islamabad 44000, Pakistan
* Correspondence: h.asadullah.1999@gmail.com
† Presented at the 2nd International Conference on Advances in Mechanical Engineering (ICAME-22), Islamabad, Pakistan, 25 August 2022.

Abstract: For any solar energy utilization, it is essential to know the value of radiation on an hourly basis and monthly basis. Generally, for Islamabad, Pakistan, available data are limited, and it is difficult to correctly evaluate the harnessed solar energy. In this paper, the monthly average solar radiation of Islamabad on an hourly basis has been presented, as compared to the available yearly average data in the literature. Moreover, the Angstrom–Prescott constant for Islamabad, Pakistan, has been calculated. The proposed relation has the error of radiation less than 1.5%.

Keywords: solar radiation; extraterritorial solar radiation; Islamabad; Pakistan; angstrom constant

check for updates

Citation: Hyder, M.J.; Nauman, A.; Musheer, S. Monthly Average Solar Radiation and Angstrom-Prescott Model for Islamabad, Pakistan. *Eng. Proc.* **2022**, *23*, 26. https://doi.org/10.3390/engproc2022023026

Academic Editors: Mahabat Khan, Muhammad Irfan and Manzar Masud

Published: 21 September 2022

Publisher's Note: MDPI stays neutral with regard to jurisdictional claims in published maps and institutional affiliations.

1. Introduction

Monthly average values of the clear sky solar radiation component for Islamabad, Pakistan, specifically for 9AM, 12PM, and 3PM, have been presented by P. Akhter et al. [1]. However, correlations have been proposed by Firaz Ahmad et al. [2] to obtain monthly average daily radiation at Karachi, Pakistan. Using various empirical relations, the mean monthly solar radiation has been estimated for Karachi, Quetta, Lahore, Multan, and Peshawar by Gadiwala M. S. [3]. Ulfat et al. have proposed the Angstrom constant for global and diffuse radiation for Islamabad [4]. G Salima et al. have determined the Angstrom constant for estimating solar radiation in Malawi [5]. S A M Maleki et al. have revisited the various solar radiation models [6].

2. Monthly Average Solar Radiation

The daily extraterrestrial solar radiation on horizontal surface (H_{oh}) is calculated from [7]:

$$H_{oh} = G_o^* t_s \left[1 + 0.033 \cos\left(360 \frac{d}{365}\right)\right] [\sin(\varnothing) \sin(\delta) + \frac{180}{\pi \omega_s} \cos(\varnothing) \cos(\delta) \sin(\omega_s)] \quad (1)$$

where $G_o{}^*$ is the solar constant ($G_o{}^*$ = 1366 w/m^2). d is the day of the year starting from 1st January, as d = 1 and φ = 33.738° is the latitude of Islamabad. δ is declination and ω_s is the hour angle at sunset.

To calculate the average monthly radiation, the midday of the month has been considered. Using Equation (1), the average monthly extraterrestrial solar radiation on a clear day of Islamabad is presented in Figure 1.

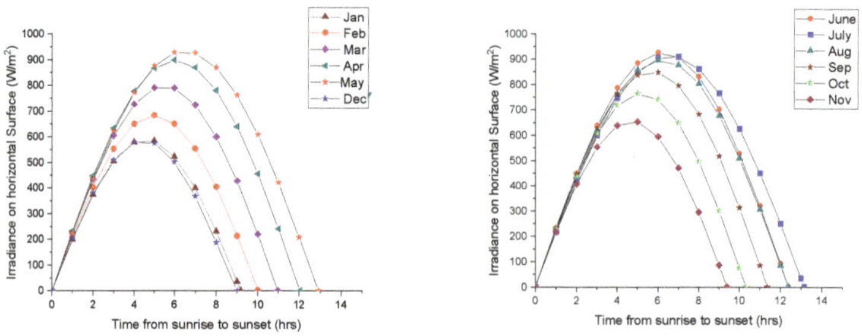

Figure 1. Monthly average solar irradiance of Islamabad.

3. Angstrom–Prescott Model for Islamabad, Pakistan

The daily solar radiation on the horizontal surface on earth (H_h) is given by the Angstrom–Prescott relation [8]:

$$\frac{H_h}{H_{oh}} = a + b\frac{n}{N} \tag{2}$$

The application of Equation (2) involves the parameters N and H_{oh} that are determined using the geographical information of the location under consideration only, where n/N is the fraction of possible sunshine, N is the maximum possible bright sunshine duration, and n is actual bright sunshine duration. The value of N is calculated from the metrological data of sunrise and sunset. It approximately takes 30 min from sunrise to reach full brightness and approximately 30 min before sunset when the brightness starts to reduce. This fact has been considered while calculating the N value. a and b are the Angstrom constants.

To evaluate the Angstrom constant (a, b), global solar radiation on the horizontal surface on earth (H_h) is taken from P. Akhter et al. [1] at 12:00 pm noon. Global extraterrestrial solar radiation on the horizontal surface (H_{oh}) is calculated using Equation (1). The actual bright sunshine duration (n) is evaluated from weather and climate [9]. Since the data pertain to the total hours of average bright sunshine of the month, the daily average value is calculated by diving by the total number of days of the month (see Table 1).

Table 1. Data for evaluating the Angstrom constant.

Month	Average Sun Shine Month	n	N	n/N	H_h	H_{oh}	H_h/H_{oh}
	(hr/Month)	(hr)	(hr)		(Global)	(Global) Calculated	(Global) Calculated
1	196	6.323	9.18	0.688	561	839	0.668
2	189	6.750	10.2	0.661	674	956	0.705
3	202	6.516	10.98	0.593	762	1134	0.671
4	249	8.300	12.05	0.688	866	1246	0.695
5	311	10.032	12.93	0.775	928	1333	0.696
6	298	9.933	13.4	0.741	897	1352	0.663
7	260	8.387	13.17	0.636	912	1302	0.700
8	249	8.032	12.38	0.648	875	1222	0.716
9	262	8.733	11.37	0.768	793	1216	0.652
10	274	8.839	10.33	0.855	690	1091	0.632
11	249	8.300	9.4	0.882	555	936	0.592
12	194	6.258	8.93	0.700	489	836	0.584

Figure 2 shows the plot of H_h/H_{oh} vs. n/N, along with the fitted curve. The Angstrom constants obtained by fitting a linear curve are $a = 0.92697 \pm 0.0485$ and $b = -0.35362 \pm 0.06104$. The Angstrom–Prescott model for Islamabad, Pakistan, is given by:

$$\frac{H_h}{H_{oh}} = 0.92697 - 0.35362\,\frac{n}{N} \tag{3}$$

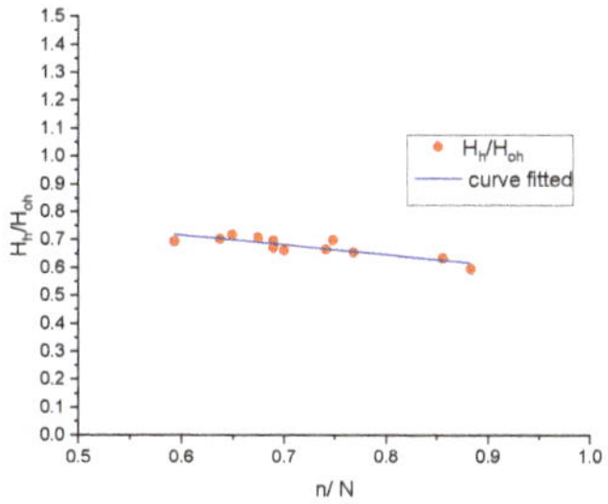

Figure 2. Linear curve fitting to obtain the Angstrom constant.

The predicted values by Equation (3) are compared with the calculated values given in Table 2. The error is less than 1.5%. Hence, the proposed model can be confidently used to calculate global solar radiation on the horizontal surface on earth at Islamabad, Pakistan.

Table 2. Comparison of calculated results (Equation (2)), predicted by Equation (3).

Month	n/N	H_h/H_{oh} (Global) Calculated	H_h/H_{oh} (Predict)	Error %
1	0.688	0.668	0.683	−1.476
2	0.661	0.705	0.692	0.012
3	0.593	0.671	0.717	−0.045
4	0.688	0.695	0.683	0.011
5	0.775	0.696	0.652	0.043
6	0.741	0.663	0.664	−0.001
7	0.636	0.700	0.701	−0.001
8	0.648	0.716	0.697	0.018
9	0.768	0.652	0.655	−0.003
10	0.855	0.632	0.624	0.008
11	0.882	0.592	0.614	−0.021
12	0.700	0.584	0.679	−0.094

Ulfat et al. [5] have proposed the Angstrom model as

$$\frac{H}{H_o} = 0.2883 + 0.5519\frac{n}{N} \tag{4}$$

The proposed model for the prediction of global solar radiation on horizontal surface by Equations (3) is compared with the Ulfat model given by Equation (4). The result is given in Table 3. It is observed that that, in some cases, the Ulfat model's error is up to 18%. Conversely, our proposed model error is less than 1.5%.

Table 3. Comparison of the proposed model with the Ulfat model.

Month	H/H_o (Global) Calculated	H/H_o Predicted	H/H_o Predicted % Error	H/H_o Ulfat	H/H_o Ulfat % Error
1	0.668	0.683	−1.476	0.668	0.024
2	0.705	0.692	0.012	0.653	5.149
3	0.671	0.717	−0.045	0.615	5.613
4	0.695	0.683	0.011	0.668	2.657
5	0.696	0.652	0.043	0.716	−2.033
6	0.663	0.664	−0.001	0.697	−3.395
7	0.700	0.701	−0.001	0.639	6.069
8	0.716	0.697	0.0185	0.646	6.966
9	0.652	0.655	−0.003	0.712	−6.007
10	0.632	0.624	0.008	0.760	−12.807
11	0.592	0.614	−0.021	0.775	−18.266
12	0.584	0.679	−0.094	0.675	−9.013

4. Conclusions

The monthly average solar radiation of Islamabad, Pakistan, is provided, and the Angstrom–Prescott model for Islamabad, Pakistan, is proposed. The proposed model predicts global solar radiation within a 1.5% error.

Author Contributions: Conceptualization, S.M. and A.N.; methodology, S.M. and M.J.H.; software, S.M. and A.N.; validation, S.M. and A.N.; formal analysis, S.M. and M.J.H.; investigation, S.M. and M.J.H.; data curation, M.J.H.; writing—original draft preparation, S.M.; writing—review and editing, S.M. and M.J.H.; supervision, M.J.H.; project administration, M.J.H. All authors have read and agreed to the published version of the manuscript.

Funding: This research received no external funding.

Institutional Review Board Statement: Not applicable.

Informed Consent Statement: Not applicable.

Data Availability Statement: Not applicable.

Conflicts of Interest: The authors declare no conflict of interest.

References

1. Akhter, P.; Baig, A.; Mufti, A. Clear Sky Solar Insolation Data for Islamabad. *Int. J. Solar Energy* **1991**, *10*, 155–159. [CrossRef]
2. Ahmad, F.; Aqil Burney, S.M.; Husain, S.A. Monthly Average Daily Global Beam and Diffuse Solar Radiation and Its Correlation with Hours of Bright Sunshine for Karachi, Pakistan. *Renew Energy* **1991**, *1*, 115–118. [CrossRef]
3. Gadiwala, M.S.; Usman, A.; Akhtar, M.; Jamil, K. Empirical Models for the Estimation of Global Solar Radiation with Sunshine Hours on Horizontal Surface in Various Cities of Pakistan. *Pak. J. Meteorol.* **2013**, *9*, 43–49.
4. Ulfat, I.; Javed, F.; Abbasi, F.A.; Kanwal, F.; Usman, A.; Jahangir, M.; Ahmed, F. Estimation of Solar Energy Potential for Islamabad, Pakistan. *Energy Procedia* **2012**, *18*, 1496–1500. [CrossRef]
5. Salima, G.; Chavula, G.M. Determining Angstrom Constants for Estimating Solar Radiation in Malawi. *Int. J. Geosci.* **2012**, *3*, 391–397. [CrossRef]
6. Mousavi Maleki, S.A.; Hizam, H.; Gomes, C. Estimation of Hourly, Daily and Monthly Global Solar Radiation on Inclined Surfaces: Models Re-Visited. *Energies* **2017**, *10*, 134. [CrossRef]
7. Twidell, J.; Weir, T. *Renewable Energy Resources*, 3rd ed.; Routledge: Abingdon-on-Thames, UK, 2015.
8. Angstrom, A. Solar and Terrestrial Radiation. *Q. J. R. Meteorol. Soc.* **1924**, *50*, 121–126. [CrossRef]
9. Weather and Climate, P. Average Monthly Hours of Sunshine in Islamabad. Available online: https://weather-and-climate.com/average-monthly-hours-Sunshine (accessed on 18 July 2022).

engineering proceedings

MDPI

Proceeding Paper

Flow Control in Paper-Based Microfluidics Using Variable Porosity Channel †

Mubashar Ali, Gohar Hussain, Hamza Abbas, Hammas Ullah and Ali Turab Jafry *

Faculty of Mechanical Engineering, Ghulam Ishaq Khan Institute of Engineering Sciences and Technology, Topi 23460, Pakistan
* Correspondence: ali.turab@giki.edu.pk
† Presented at the 2nd International Conference on Advances in Mechanical Engineering (ICAME-22), Islamabad, Pakistan, 25 August 2022.

Abstract: Microfluidic paper-based analytical devices have broadened the scope of microfluidics by offering low-cost, simple, faster fabrication, biodegradability, and environmentally benign devices capable of medical diagnostics, environmental sensing, and food quality control. These devices are incredibly versatile, which is one of their most notable features. Despite recent advancements in paper-fluidics, creating slow flow channels in paper still remains a challenge. Herein, we propose viscous barriers of various concentrations embedded along the paper channel to control its flow velocity by altering the pore size. We used sugar concentrations in the range of 0–40% dried in porous media and then recorded flow behaviors of water and castor oil. From experiments, it was observed that by increasing the sugar concentration, delay time also increased. Moreover, changing the type of fluid (w.r.t viscosity) also varies the flow delays as castor oil took a much longer time to cover the same channel length as compared to water. We believe that our proposed method will play an important part in improving flow delays and can be applied for food quality applications such as time–temperature indicators due to its simple fabrication and cost-effective technique.

Keywords: delay zone; food quality; microfluidic; paper-based microfluidics

Citation: Ali, M.; Hussain, G.; Abbas, H.; Ullah, H.; Jafry, A.T. Flow Control in Paper-Based Microfluidics Using Variable Porosity Channel. *Eng. Proc.* **2022**, *23*, 33. https://doi.org/10.3390/engproc2022023033

Academic Editors: Mahabat Khan, M. Javed Hyder, Muhammad Irfan and Manzar Masud

Published: 26 September 2022

1. Introduction

Global health is the most important challenge that the world faces in improving disease detection and diagnosis precision. Health can be improved by providing an adequate quality of food and a pollution-free environment. Microfluidics devices provide great potential for food safety and inexpensive diagnostics. These devices enable less sample consumption as well as quicker reaction and separation times. Integrated microfluidic devices are used for a wide range of applications. These applications include medical diagnostics, environmental sensing, drug discovery, drug delivery, chemical and biochemical processes in the field of analysis, as well as energy conversion and storage [1–5]. Such devices are mainly used for immunosensing, point of care (POC), and filtration of biological fluid on a chip [6]. Martinez et al. of Whiteside's group at Harvard realized the use of paper for microfluidic devices in 2007 and reported first microfluidic paper-based analytical device (μPAD) [6]. The developing cost of μPADs is minimal, because paper is the cheaper, flexible, portable, and biodegradable material. As a result, there is no external force required to control the flow of the fluid in the paper due to its capillary force.

Understanding of flow control is very important in paper-based microfluidic devices for its accuracy and predictability [7,8]. Flow control in a two-dimensional paper-based microfluidics device can be done by either changing the geometry or by changing the chemical properties of the flowing fluid [9]. A simple geometric control can be done by just changing the channel length to control wicking times [10]. Printing baffles on the paper can also be used to decrease flow time [11]. Chemical flow control can be done by using sugar and wax to create barriers in the path of the flowing fluid to decrease flow time for food quality control.

For our research, we have used sucrose for creating barriers in the path of the flowing fluid. By adding sugar barriers into the paper channel, porosity will decrease, and as a result, imbibition of the solution into the paper will decrease. Decrease in the pore size of the paper will increase the resistance of the flowing fluid in the paper channel; as the resistance is increased, the flowing fluid will take more time to reach the required point. In this research, we adopted the chemical-based method to control and manipulate fluid flow in paper for food quality devices such as time–temperature indicators to increase the delay time of fluid flow in paper. The flow control was achieved by changing the amount of sucrose in the solution of wetted paper.

2. Theoretical Background

Flow can be classified as wet-out flow and fully-wetted flow. One-dimensional fluid flow into a dry porous paper channel is termed as wet-out flow. The classic Lucas–Washburn (L–W) concept of wet-out in a paper pore does not take part in the complicated reduction process of sugar-treated strips; it would provide a useful framework for describing the origin of time delays. The L–W equation is derived from the capillary force due to surface tension of the paper and the viscous force of fluid flow. It explains that the length of liquid imbibition is proportional to the square root of time.

$$L^2 = \frac{r\gamma cos\theta}{4\mu}t \tag{1}$$

where L is the penetration length of the wicking fluid, r is the pore size radius of the paper strip, γ is the surface tension force, θ is the contact angle and μ is the dynamic viscosity of the flowing fluid. The capillary force responsible for fluid flow in paper is caused by the surface tension γ at the fluid surface. The higher the capillary pull, the higher the surface tension at the liquid interface, and hence the higher the fluid flow in the paper channel.

3. Materials and Methods

The delay time of wet-out flow through the paper strip is increased by creating the delay zone of different sugar concentrations. A sugar solution is prepared by mixing the sucrose in distilled water at room temperature. Solubility of sucrose in water at 200 °C is 66.7% by mass. We prepared the sugar solution in distilled water of 30% and 40% sucrose by mass. Whatman filter paper strips of grade 40 were inserted in 30% and 40% sugar solution until the solution was wicked at the desired point on the strip. Wicking strips were dried at room temperature. Figure 1 shows the preparation of the sugar delay zone of different concentrations.

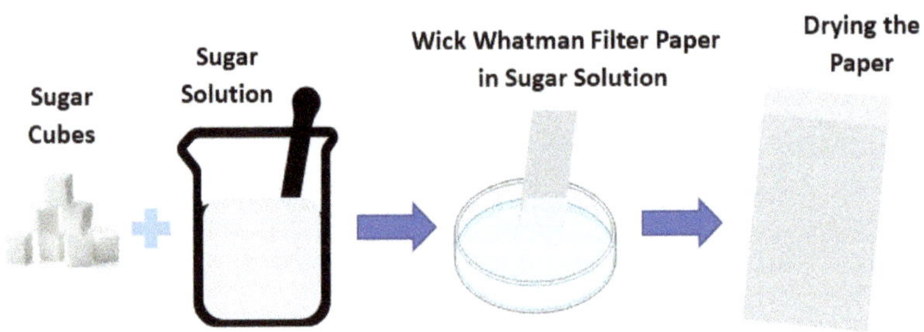

Figure 1. Preparation of sugar delay zone.

As the length travelled by liquid in the rectangular channel increases, viscous drag acting on it also increases, which causes the flow velocity to decrease with time. Figure 2 shows the image during the experiment. The flow speed of the water is high in the low concentration delay zone. For the high concentration sugar delay zone, the liquid flow is slow. Time to reach the fluid at the finish line is increased by increasing the sugar concentration from 0–40%. The length of liquid flow in the paper strip is measured by the Image J software (National Institutes of Health and LOCI, 1.48 v, Bethesda, MD, USA, and Madison, WI, USA), as shown in Figure 2c.

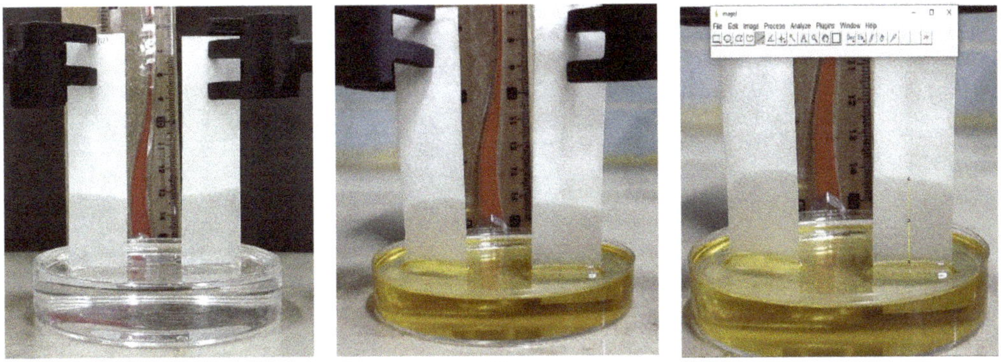

(a) (b) (c)

Figure 2. Image during experiment: (**a**) Water flow untreated strip; (**b**) flow of castor oil in 30% sugar concentration strip; (**c**) penetration length measured with Image J software (1.48 v).

4. Results and Discussion

Figure 3a shows the flow of water in the paper strips of different sugar concentrations. Water flows in the untreated 0% sugar concentration strip for 7 min to reach a length of 60 mm. At the creation of a delay zone of 30% sugar concentration, it takes 18 min to reach the finish line of the 60 mm length. Water flowing in a created delay zone of 40% sugar concentration, on the other hand, takes 20 min to reach a length of 28 mm. So, this indicates that wicking delay time can be increased by increasing the sugar concentration to some extent.

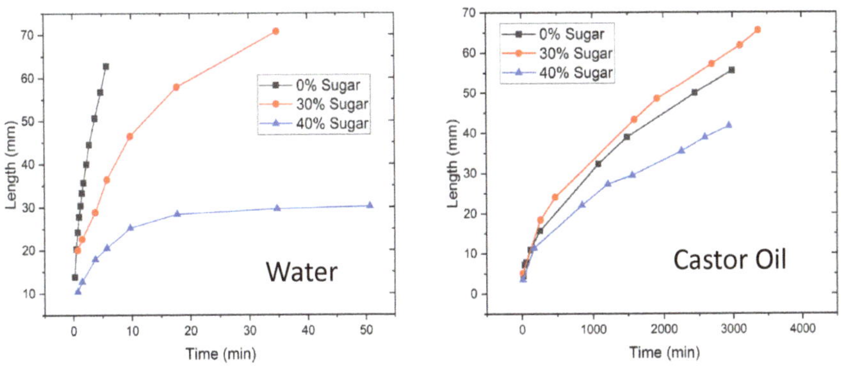

(a) (b)

Figure 3. Length–time curve: (**a**) Flow of water in sugar delay zone of different concentrations; (**b**) flow of castor oil in sugar delay zone of different concentrations.

To increase the delay time from hours to days, we used the high viscous fluid of castor oil, which has a viscosity of 14 times more than water. The flow of castor oil in the untreated paper strip of 0% sugar solution takes two and a half days to reach the same length of 60 mm as shown in Figure 3b. However, in the delay zone of 40% sugar concentration, castor oil takes much more time, 2 days, to cover the length of 40 mm. Hence, the results show that time delay can be increased by creating a sugar delay zone of different concentrations and changing the wicking fluid from water to a high viscosity liquid.

5. Conclusions

Flow delays were achieved by using sucrose solutions of various concentrations in paper media to study their effect on flow delays. Several experiments were performed by changing sucrose concentrations in distilled water in a range of 0–40%. Moreover, the flow behavior of highly viscous fluid, e.g., castor oil, was also studied and compared with distilled water. From experiments, it was recorded that a flow delay of 40 min was achieved by using a 40% sucrose concentration. To achieve a flow delay of two days, castor oil is also promising in this regard. We believe that this study will help researchers to develop more sustainable and efficient methods of flow delays for food quality and environmental sensing applications of paper-based microfluidic devices.

Author Contributions: Conceptualization, M.A. and G.H.; methodology, M.A. and A.T.J.; software, M.A. and H.A.; validation, M.A. and H.A.; formal analysis, M.A. and H.U.; investigation, M.A. and A.T.J.; data curation, H.U.; writing—original draft preparation, M.A.; writing—review and editing, M.A. and A.T.J.; supervision, A.T.J. All authors have read and agreed to the published version of the manuscript.

Funding: This work was funded by Higher Education Commission (HEC), Pakistan under the Technology Transfer Support Fund (grant no. TTSF-74).

Institutional Review Board Statement: Not applicable.

Informed Consent Statement: Not applicable.

Data Availability Statement: Not applicable.

Conflicts of Interest: The authors declare no conflict of interest.

References

1. Dittrich, P.S.; Manz, A. Lab-on-a-Chip: Microfluidics in Drug Discovery. *Nat. Rev. Drug Discov.* **2006**, *5*, 210–218. [CrossRef] [PubMed]
2. Aryasomayajula, A.; Bayat, P.; Rezai, P.; Selvaganapathy, P.R. Microfluidic Devices and Their Applications. In *Springer Handbook of Nanotechnology*; Bhushan, B., Ed.; Springer: Berlin/Heidelberg, Germany, 2017; pp. 487–536. ISBN 978-3-662-54357-3.
3. Safdar, M.; Jänis, J.; Sánchez, S. Microfluidic Fuel Cells for Energy Generation. *Lab Chip* **2016**, *16*, 2754–2758. [CrossRef] [PubMed]
4. Nguyen, N.-T.; Wereley, S.T.; House, A. *Fundamentals and Applications of Microfluidics*, 2nd ed.; Artech House: Boston, MA, USA, 2002.
5. Khan, I.U.; Serra, C.A.; Anton, N.; Vandamme, T.F. Production of Nanoparticle Drug Delivery Systems with Microfluidics Tools. *Expert. Opin. Drug Deliv.* **2015**, *12*, 547–562. [CrossRef] [PubMed]
6. Li, X.; Chen, W.; Liu, G.; Lu, W.; Fu, J. Continuous-Flow Microfluidic Blood Cell Sorting for Unprocessed Whole Blood Using Surface-Micromachined Microfiltration Membranes. *Lab Chip* **2014**, *14*, 2565–2575. [CrossRef] [PubMed]
7. Martinez, A.W.; Phillips, S.T.; Butte, M.J.; Whitesides, G.M. Patterned Paper as a Platform for Inexpensive, Low-Volume, Portable Bioassays. *Angew. Chem.-Int. Ed.* **2007**, *46*, 1318–1320. [CrossRef] [PubMed]
8. Lim, H.; Jafry, A.T.; Lee, J. Fabrication, Flow Control, and Applications of Microfluidic Paper-Based Analytical Devices. *Molecules* **2019**, *24*, 2869. [CrossRef] [PubMed]
9. Fu, E.; Lutz, B.; Kauffman, P.; Yager, P. Controlled Reagent Transport in Disposable 2D Paper Networks. *Lab Chip* **2010**, *10*, 918–920. [CrossRef] [PubMed]
10. Whitesides, G.M. Viewpoint on "Dissolvable Fluidic Time Delays for Programming Multi-Step Assays in Instrument-Free Paper Diagnostics". *Lab Chip* **2013**, *13*, 4004–4005. [CrossRef] [PubMed]
11. Preechakasedkit, P.; Siangproh, W.; Khongchareonporn, N.; Ngamrojanavanich, N.; Chailapakul, O. Development of an Automated Wax-Printed Paper-Based Lateral Flow Device for Alpha-Fetoprotein Enzyme-Linked Immunosorbent Assay. *Biosens. Bioelectron.* **2018**, *102*, 27–32. [CrossRef] [PubMed]

engineering proceedings

Proceeding Paper

Numerical Investigation of Impact and Compression after Impact Performance of 45° Biaxial Composite Laminates †

Manzar Masud *, **Muhammad Daud Ali**, **Muhammad Irfan**, **Shummaila Rasheed** and **Muhammad Haroon**

Department of Mechanical Engineering, Capital University of Science and Technology (CUST), Islamabad 44000, Pakistan
* Correspondence: manzar.masud@cust.edu.pk
† Presented at the 2nd International Conference on Advances in Mechanical Engineering (ICAME-22), Islamabad, Pakistan, 25 August 2022.

Abstract: A meso-micro analysis technique established on the basis of micro-mechanics of failure in combination with progressive-based damage criteria of the composite material constituents (i.e., fiber and matrix) is demonstrated to predict the impact and compression after impact (CAI) performance of biaxial composite laminates. Damages in the composite material constituents are calculated using different failure models. The analysis technique is then used to investigate the impact and CAI behavior of 45° biaxial composite laminates, for both thermoset and thermoplastic resin systems. The results were presented in the form of damage propagation contours for both impact and compression after impact and a comparison of graphs showing displacement-time, internal energy-time, velocity-time, and contact force-time for both thermoset and thermoplastic resin.

Keywords: multi-scale; micromechanics of failure; impact; compression after impact; biaxial composite laminates

check for updates

Citation: Masud, M.; Ali, M.D.; Irfan, M.; Rasheed, S.; Haroon, M. Numerical Investigation of Impact and Compression after Impact Performance of 45° Biaxial Composite Laminates. *Eng. Proc.* **2022**, *23*, 18. https://doi.org/10.3390/engproc2022023018

Academic Editors: Mahabat Khan and M. Javed Hyder

Published: 20 September 2022

Publisher's Note: MDPI stays neutral with regard to jurisdictional claims in published maps and institutional affiliations.

1. Introduction

Recently, an exceeding growth of the application of biaxial composite laminates have been observed due to a high ratio of strength to weight, an ability to tailor the material for different applications and resistance to damage due to impact load [1]. Due to the recent advancements in the aircraft-based industry and increased claim for the composite laminate based structures, the demand of accurate failure prediction models has also increased.

For the design and development of structures based on composite materials, the numerical investigation can play a vital role in the reduction in resources such as cost and time. For accurate numerical analysis and prediction of complex failure phenomenon due to impact, dependable failure theories and models based on progressive damage are required to accurately investigate the failure mechanism in the composite laminates-based structures [2]. Therefore, predictive tools requiring a smaller quantity of essential mechanical tests are becoming more significant.

The failure models dependent on the micromechanics technique have exceedingly developed expediency for investigators by accurately developing the complex mechanisms in the composite laminate based structures; having an advantage in investigating the failure prediction at constituents-based level [3]. Due to the combination of meso-level and micro-level representative unit cells, this multi-scale technique takes the accuracy and efficiency of both micromechanics and macro-based analysis.

A meso-micro analysis technique established on the basis of micromechanics of failure (MMF) technique [4] and progressive based damage analysis is used to predict the impact and compression after impact (CAI) performance of biaxial (BX) laminates. MMF is dependent on progressive damage models for constituents, i.e., fiber and matrix, in which there is a transfer of data from micro-scale to meso-scale and back again. In view of the advantage of accurate MMF predictions [4], MMF based technique is applied to predict the impact and CAI behavior of BX 45 laminates made with a thermoset and thermoplastic resin system.

2. Methodology

For impact analysis, an explicit model-based scheme was used to perform the simulation of the impact loading using ABAQUS 6.14 (Dassault Systèmes, Vélizy-Villacoublay, France), where dynamic explicit solver was used. An algorithm was established for combining MMF and a progressive damage model and was implemented using VUSDFLD.

In this study, two types of resin systems, i.e., thermoset and thermoplastic were investigated for BX 45 composite laminate. The used fiber (T700) behaves as transversely isotropic material, having five independent material properties [5], whereas the matrix behaves isotropically, having two material properties [5].

The geometry, dimensions and the boundary conditions are given in Figure 1a. The ply thickness for all investigated models was taken as 0.125 mm. The radius of the impactor model was taken as 5 mm. The starting location of the impactor was established above the BX laminate model. Four sides of the composite laminate were constrained using fixed boundary conditions. After the impact analysis, CAI was performed, in which the left-over damages from impact analysis were introduced to the laminate and a compression test was performed. The procedure showing the methodology is shown in Figure 1b.

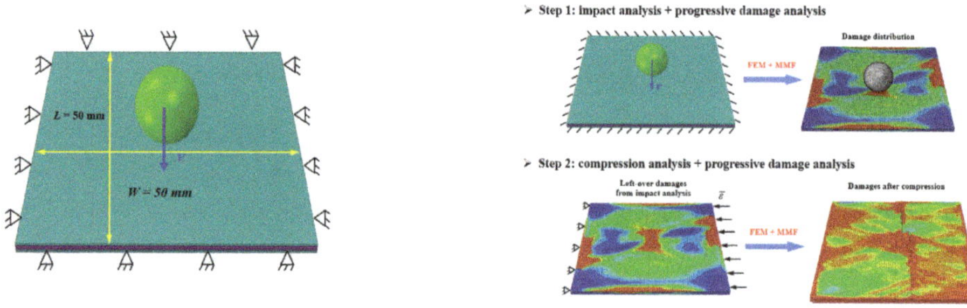

(a) (b)

Figure 1. (a) Geometry and Boundary Conditions (b) Methodology adopted for impact and CAI.

In the developed FE geometries, the plies were made with the help of an eight-node linear brick element, i.e., C3D8R. [6]. Mesh sensitivity analysis, which is an important step in the simulation process, was performed and it was concluded that the selected mesh remained a compromise amid the quality of output and computational time.

3. Results and Discussion

The impact resistance and CAI behavior of the BX 45 composite laminate made with thermoset and thermoplastic resins were quantitatively assessed by a visual examination of the contours induced by damage, as they are exposed to an impact of 50 m/s. The Figure 2a–f shows the damage contours of impact and CAI for the BX 45 composite laminate along with the displacement, velocity, internal energy and contact force time histories for V = 50 m/s.

The results shown in Figure 2a are the ply-by-ply damage contours, indicating the damage propagation for the impact case. The damage contours are shown at the time step of t = 0.05 s and t = 0.1 s. As a result of the visual examination, it is qualitatively concluded, due to the use of thermoplastic resin, that BX 45 laminates caused an improved dispersal of the impact-based energy; a lesser damage area due to deletion of elements, and, therefore, an enhanced resistance due to impact.

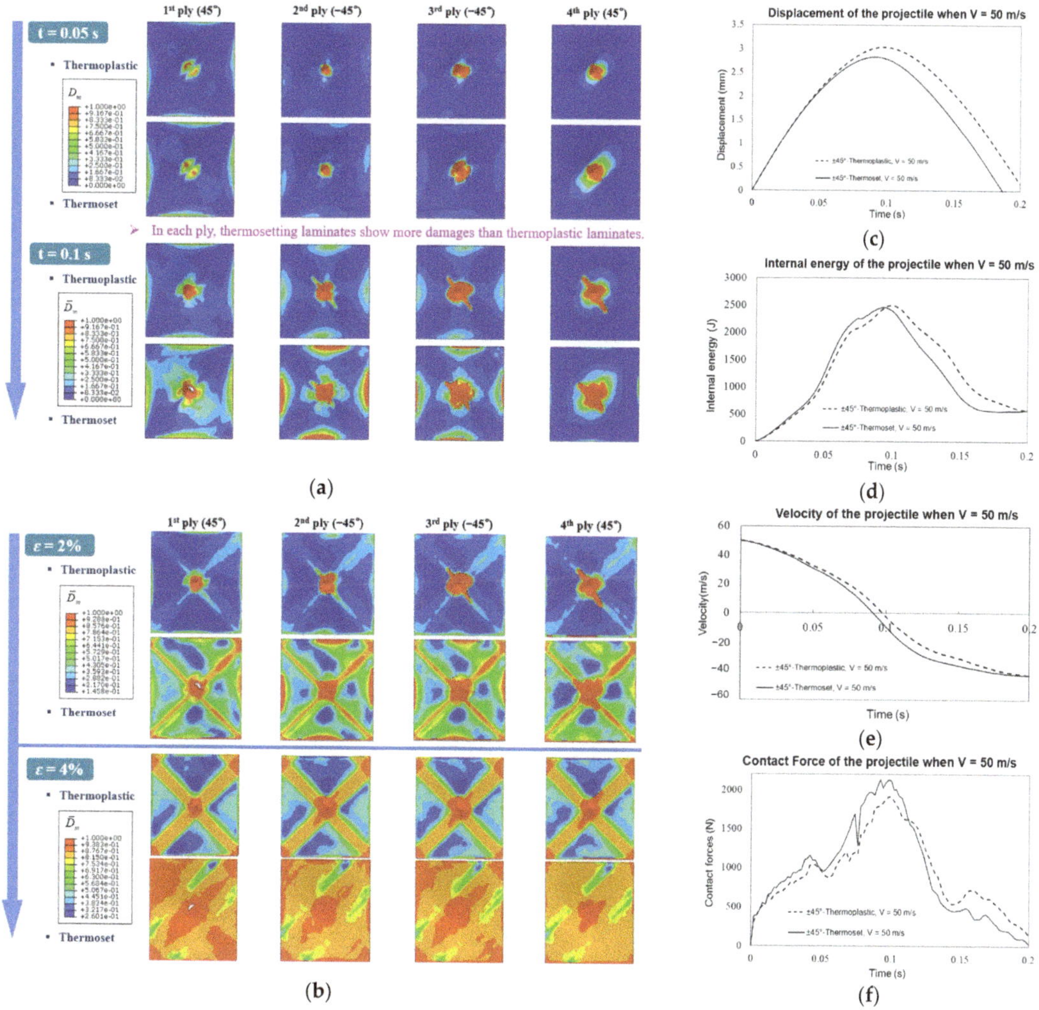

Figure 2. (**a**) Damage contours for Impact; (**b**) damage contours for CAI; (**c**) Displacement vs. time graph; (**d**) internal energy vs. time graph; (**e**) velocity vs. time graph (**f**) contact force vs. time graph for BX 45 composite laminates.

Figure 2b shows the results for the CAI case, in which the damaged biaxial composite laminate due to impact is subjected to compression strain of 2% and 4%. The damage contours for the CAI case shows that less damage is induced in the thermoplastic case as compared to the thermosetting case, thus indicating there is more residual strength in the case of thermoplastic resins based biaxial composite laminates.

To further investigate resistance due to the impact of BX 45composite laminate, the assessment of absorption of impact energy was performed and results are shown in Figure 2c–f. The evaluation includes the investigation of reduction in displacement and velocity of projectile after impact. The absorption capacity of impact energy for composite laminates is investigated by analyzing the decrease in residual velocities against the incident impact velocity. The results show that the decrease in residual velocity was higher in the case of thermosetting resin than the thermoplastic reason. The higher the decrease in residual velocity of the projectile, the higher the impact energy absorption. On the basis of the assessment of impact damaged areas, it can be deduced that BX 45 composite modeled using thermoplastic resin shows enhanced resistance due to impact compared to the BX 45 composite modeled using thermosetting resin.

4. Conclusions

In this research, a multi-scale technique dependent on MMF in combination with a progressive damage model is demonstrated, and later applied on the 45-degree biaxial composite laminate, made of thermosetting and thermoplastic resin system. The full penetration of an impactor with a velocity of 50 m/s was simulated, followed by a compression after impact simulation. The impact resistance was assessed based on the quantitative analysis using damage contours for both impact and CAI and also on the basis of impact energy absorption. Results showed that the BX 45 composite laminates with thermoplastic resin results in enhanced resistance due to impact compared to composite laminates with thermosetting resin. The demonstrated methodology for investigating the impact resistance of composite laminates is relatively general as it initiates with the vital composite material's constituents, fiber and matrix. As soon as the material properties are attained, the proposed technique can be used to investigate any material and structure, subjected to different types of loadings.

Author Contributions: Conceptualization, M.M. and M.D.A.; methodology, M.M. and M.I.; software, M.M. and S.R.; validation, M.M. and S.R.; formal analysis, M.M. and M.H.; investigation, M.M. and M.I.; data curation, M.H.; writing—original draft preparation, M.M.; writing—review and editing, M.M. and M.I.; supervision, M.I.; project administration, M.M. All authors have read and agreed to the published version of the manuscript.

Funding: This research received no external funding.

Institutional Review Board Statement: Not applicable.

Informed Consent Statement: Not applicable.

Data Availability Statement: Not applicable.

Acknowledgments: The authors would like to acknowledge the support of the Department of Mechanical Engineering, Capital University of Science and Technology, Islamabad.

Conflicts of Interest: The authors declare no conflict of interest.

References

1. Jefferson, A.J.; Arumugam, V.; Dhakal, H.N. 1—Introduction and Context. In *Repair of Polymer Composites*; Jefferson, A.J., Arumugam, V., Dhakal, H.N., Eds.; Woodhead Publishing Series in Composites Science and Engineering; Woodhead Publishing: Sawston, UK, 2018; pp. 1–43, ISBN 978-0-08-102263-4.
2. Shah, S.Z.H.; Karuppanan, S.; Megat-Yusoff, P.S.M.; Sajid, Z. Impact Resistance and Damage Tolerance of Fiber Reinforced Composites: A Review. *Compos. Struct.* **2019**, *217*, 100–121. [CrossRef]
3. Azeem, M.; Irfan, M.; Masud, M.; Rehman, G.U.; Ali, H.; Ali, M.U.; Zafar, A.; Muhammad Niazi, U.; Rahman, S.; Legutko, S.; et al. Experimental and Numerical Investigation of Effect of Static and Fatigue Loading on Behavior of Different Double Strap Adhesive Joint Configurations in Fiber Metal Laminates. *Materials* **2022**, *15*, 1840. [CrossRef] [PubMed]
4. Masud, M.; Kharusi, M.S.M.A.; Ali, M.U.; Mubashar, A.; Hussain, S.J.; Tariq, A.; Rehman, G.U.; Akhtar, M.H.; Javeed, S. Prediction of the Ultimate Strength of Notched and Unnotched IM7/977-3 Laminated Composites Using a Micromechanics Approach. *Polymers* **2021**, *13*, 3491. [CrossRef] [PubMed]
5. Lei, X.; Hayat, K.; Ha, S.K. A Multi-Scale Approach to Predict the Impact Resistance of Braided Composites Using Progress Failure. *Adv. Compos. Lett.* **2017**, *26*, 24–29. [CrossRef]
6. Systèmes, D. Abaqus Standard and Abaqus Documentation for Version 6.14. *Dassault Syst. Simulia Corp.* **2014**, *651*, 2–6.

engineering proceedings

MDPI

Proceeding Paper

Cellulose Triacetate/Zinc Oxide Membrane for Bioethanol Recovery via Pervaporation †

Asif Shahzad [1], Meshal Alzaid [2,*], Muhammad Ammar [3], Rizwan Ahmed Malik [4] and Muddassir Ali [5,*]

1 Department of Mechanical Engineering, Faculty of Mechanical and Aeronautical Engineering, University of Engineering and Technology, Taxila 47080, Pakistan
2 Department of Physics, College of Science, Jouf University, Sakaka 72388, Saudi Arabia
3 Department of Chemical Engineering Technology, Faculty of Engineering, Government College University, Faisalabad 38000, Pakistan
4 Department of Metallurgy and Materials Engineering, Faculty of Mechanical and Aeronautical Engineering, University of Engineering and Technology, Taxila 47080, Pakistan
5 Department of Energy Engineering, Faculty of Mechanical and Aeronautical Engineering, University of Engineering and Technology, Taxila 47080, Pakistan
* Correspondence: mmalzaid@ju.edu.sa (M.A.); muddassir.ali@uettaxila.edu.pk (M.A.)
† Presented at the 2nd International Conference on Advances in Mechanical Engineering (ICAME-22), Islamabad, Pakistan, 25 August 2022.

Abstract: In this study, cellulose triacetate (CTA) hybrid membrane is successfully prepared via the phase-inversion method for bioethanol recovery through pervaporation. Nano zinc oxide (ZnO) particles are mixed into the polymer matrices of CTA to enhance the pervaporation membrane's performance. The fabricated hybrid membrane is characterized using environmental scanning electron microscopy (ESEM) and thermogravimetric analysis (TGA) to reveal the surface morphology and thermal resistance, respectively. The pervaporation performance of the hybrid membrane is assessed for recovering bioethanol from its dilute solution. Pervaporation results show that the hybrid membrane prepared with 3 wt.% ZnO achieved a permeation flux of 1065.71 g/m^2h, while the separation factor was around 1038 at 50 °C operating temperature.

Keywords: bioethanol; cellulose triacetate; mixed-matrix membrane; pervaporation; zinc oxide (ZnO)

check for updates

Citation: Shahzad, A.; Alzaid, M.; Ammar, M.; Malik, R.A.; Ali, M. Cellulose Triacetate/Zinc Oxide Membrane for Bioethanol Recovery via Pervaporation. *Eng. Proc.* **2022**, 23, 31. https://doi.org/10.3390/engproc2022023031

Academic Editors: Mahabat Khan, M. Javed Hyder, Muhammad Irfan and Manzar Masud

Published: 23 September 2022

1. Introduction

Fortunately, the use of renewable energy resources and environmentally friendly production methods is rising, and biofuels are emerging as a viable alternative to fossil fuels [1]. Bioethanol has been used in the last few years as a substitute for gasoline owing to its low-cost, high-octane rating, and low environmental impact [2]. However, the profitability of bioethanol is considerably reduced due to the difficulties involved in its production and separation. One of the significant bottlenecks among them is product inhibition during the fermentation stage, which minimizes the ethanol titer and productivity [3]. Accordingly, in situ ethanol recovery from fermentation broth can improve ethanol yield and productivity. Recently, pervaporation has been demonstrated to be beneficial for in situ ethanol recovery, which, in turn, enhances product inhibition in the fermenter [4]. Pervaporation is an appealing membrane-based technique owing to its enhanced efficiency, separation selectivity, low energy consumption, and lack of toxicity to microorganisms [5].

The membrane material plays a significant role in the fabrication of pervaporation membranes. It has been seen that combining polymeric and inorganic nanofillers to create hybrid membranes is an easy and cost-effective approach to increase membrane selectivity and overcome trade-off limitations [6,7]. Incorporating suitable nanofillers could improve pervaporation performance by enhancing selective diffusion and sorption, or both. A cellulose derivative, cellulose triacetate (CTA), has gained prominence again as a membrane material because of its superior mechanical characteristics, less fouling affinity, better hydrophilicity, and excellent oxidation stability. Moreover, it can be fabricated into a very dense film, which is requisite for use as PV membranes [8]. As a low-cost, multifunctional inorganic nanofiller, zinc oxide (ZnO) has lately received much attention as a result of its unique chemical and physical properties, such as superior stability, nontoxic nature, and potent bactericide and antibacterial characteristics [9]. Moreover, ZnO nanofiller's better hydrophilicity has shaped it as a suitable filler for enhancing the hydrophilicity of hybrid membranes [10]. Accordingly, mixing ZnO nanofillers into the CTA matrix for pervaporation is a promising strategy for recovering bioethanol from fermentation broths.

This study employs the phase inversion technique to fabricate a hybrid membrane consisting of cellulose triacetate polymer and ZnO nanofillers. The effects of nano-ZnO fillers on the physicochemical characteristics of the hybrid membrane, such as structural morphology and thermal properties, are also assessed during this study. Furthermore, the PV performance of the fabricated hybrid membrane is evaluated during the recovery of bioethanol from the aqueous solution as a function of feed content.

2. Materials and Methods

Membrane Preparation, Characterization, and Performance

CTA–ZnO hybrid membrane is fabricated via the phase inversion method. First, 3 wt.% ZnO nanoparticles are dispersed in dimethyl sulfoxide (DMSO) solvent for 1 h in an ultrasonic bath operating at 60 °C. Then, the suspension solution of ZnO is stirred for 2 h at 75 °C. Next, CTA solid polymer is added to the suspension of ZnO and stirred for 6 h at 75 °C to obtain a uniform dope solution. The dope solution is placed for 12 h in the fume hood to eradicate air bubbles formed during the stirring process. Then, a fixed amount of dope solution is poured into Petri dishes and left inside the fume hood for 48 h to remove the solvent properly via evaporation at room temperature. Finally, the dried membrane is detached from the Petri dishes and washed with deionized water to remove the remaining solvent.

Morphological assessment of the fabricated hybrid membrane is examined using environmental scanning electron microscopy (ESEM), Thermo Fisher Scientific, Waltham, MS, USA. A thermogravimetric analyzer (TGA), Shimadzu, Kyoto, Japan is used to examine the thermal characteristics of the fabricated hybrid membrane. The pervaporation performance of the CTA–ZnO hybrid membrane in the recovery of bioethanol from the aqueous solution is investigated using a locally fabricated experimental setup.

3. Results and Discussion

3.1. Membrane Characterization

The influence of 3 wt.% ZnO nanoparticles on the morphological characteristics of hybrid CTA–ZnO membrane is assessed by SEM analyses and shown in Figure 1. The CTA–ZnO fabricated hybrid membrane cross-sectional morphology depicts a sponge-like porous structure in Figure 1a. It can be seen that mixing ZnO nanofillers into the CTA matrix results in a surface made of numerous protrusions, as shown in Figure 1b. SEM analysis indicates that CTA membrane mixed with ZnO nanofillers results in a porous and asymmetrical structural membrane.

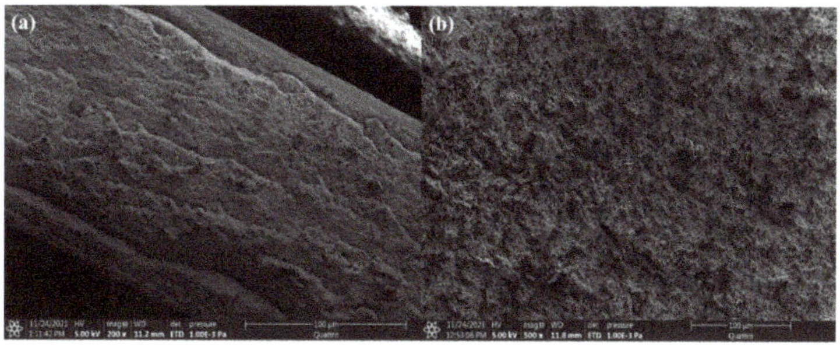

Figure 1. Morphology of hybrid membrane: (**a**) cross section; (**b**) surface.

TGA analysis is carried out to assess the thermal characteristics of fabricated hybrid CTA–ZnO membrane. Figure 2 demonstrates that the thermal degradation of the hybrid membrane occurs in three stages. It can be seen that the hybrid membrane started to degrade at 95 °C and continued to degrade up to 380 °C. The addition of ZnO nanofillers considerably altered the thermal resistance of the fabricated hybrid membrane.

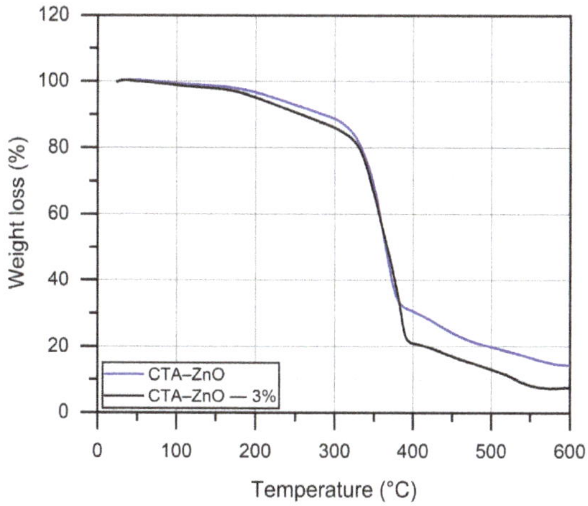

Figure 2. TGA analysis of hybrid membranes.

3.2. *Pervaporation Performance*

The effect of the water content in the feed on the pervaporation performance of the CTA–ZnO hybrid membrane at 50 °C is shown in Figure 3. Results indicate that the permeation flux and the separation factor strongly depend on the feed content. It can be seen that, with increasing water content in the feed, the permeation flux increases considerably because of the improved membrane affinity towards water. In addition, increasing from 10 wt.% to 25 wt.%, the water concentration in the feed reduces the separation factor substantially because of better membrane swelling. Finally, the maximum permeation flux of 1065.71 g/m^2h at the concentration of 25 wt.%, while the highest separation factor is around 1038 at 10 wt.% feed concentration.

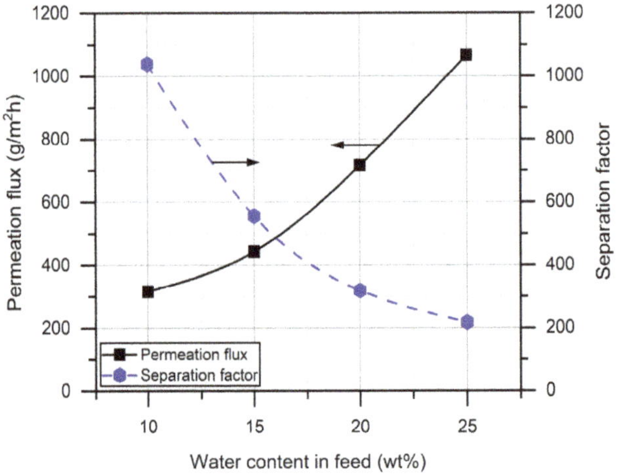

Figure 3. Pervaporation performance of hybrid membrane for water concentration in feed at 50 °C.

4. Conclusions

A phase inversion approach is employed to fabricate CTA–ZnO hybrid membrane for bioethanol recovery via pervaporation. It is revealed that adding 3 wt.% ZnO nanofillers to the CTA polymer matrix enhanced the permeation flux. Furthermore, the addition of ZnO nanofillers also enhanced the membranes' chemical composition, hydrophilicity, and thermal stability. The maximum permeation flux and separation factor achieved by the hybrid membrane at an operating temperature of 50 °C is 1065.71 g/m^2h and 1038, respectively.

Author Contributions: Conceptualization, R.A.M. and M.A. (Muddassir Ali); methodology, M.A. (Muddassir Ali) and A.S.; validation, A.S., M.A. (Meshal Alzaid) and M.A. (Muhammad Ammar); formal analysis, A.S.; investigation, M.A. (Muddassir Ali) and A.S.; resources, M.A. (Muddassir Ali); data curation, A.S.; writing—original draft preparation, A.S.; writing—review and editing, M.A. (Muddassir Ali); supervision, M.A. (Muddassir Ali); project administration, M.A. (Muddassir Ali) and M.A. (Meshal Alzaid); funding acquisition, M.A. (Meshal Alzaid). All authors have read and agreed to the published version of the manuscript.

Funding: This research work is sponsored by the Deanship of Scientific Research of Jouf University, Saudi Arabia (DSR-2021-03-03108).

Institutional Review Board Statement: Not applicable.

Informed Consent Statement: Not applicable.

Acknowledgments: The authors appreciate the technical support provided by the Department of Mechanical Engineering, University of Engineering and Technology, Taxila–Pakistan, with this research work.

Conflicts of Interest: The authors declare no conflict of interest.

References

1. Karimi, S.; Karri, R.R.; Yaraki, M.T.; Koduru, J.R. Processes and separation technologies for the production of fuel-grade bioethanol: A review. *Environ. Chem. Lett.* **2021**, *19*, 2873–2890. [CrossRef]
2. Hajilary, N.; Rezakazemi, M.; Shirazian, S. Biofuel types and membrane separation. *Environ. Chem. Lett.* **2019**, *17*, 1–18. [CrossRef]
3. Knozowska, K.; Kujawska, A.; Li, G.; Kujawa, J.; Bryjak, M.; Kujawski, W.; Lipnizki, F.; Ahrné, L.; Petrinić, I.; Kujawski, J.K. Membrane assisted processing of acetone, butanol, and ethanol (ABE) aqueous streams. *Chem. Eng. Process. Process Intensif.* **2021**, *166*, 108462. [CrossRef]
4. Aghaei, S.; Alavijeh, M.K.; Shafiei, M.; Karimi, K. A comprehensive review on bioethanol production from corn stover: Worldwide potential, environmental importance, and perspectives. *Biomass Bioenergy* **2022**, *161*, 106447. [CrossRef]

5. Zheng, P.; Li, C.; Wang, N.; Li, J.; An, Q. The potential of pervaporation for biofuel recovery from fermentation: An energy consumption point of view. *Chin. J. Chem. Eng.* **2019**, *27*, 1296–1306. [CrossRef]
6. Peng, P.; Lan, Y.; Liang, L.; Jia, K. Membranes for bioethanol production by pervaporation. *Biotechnol. Biofuels* **2021**, *14*, 10. [CrossRef] [PubMed]
7. Cai, W.; Cheng, X.; Chen, X.; Li, J.; Pei, J. Poly(vinyl alcohol)-modified membranes by $Ti_3C_2T_x$ for ethanol dehydration via pervaporation. *ACS Omega* **2020**, *5*, 6277–6287. [CrossRef] [PubMed]
8. Fareed, J.; Zafar, M.; Saleem, M.; Malik, R.A.; Ali, M. Cellulose Blended Membranes for High-Salinity Water Pervaporation Desalination. *Eng. Proc.* **2021**, *12*, 57. [CrossRef]
9. Asiri, A.M.; Petrosino, F.; Pugliese, V.; Khan, S.B.; Alamry, K.A.; Alfifi, S.Y.; Marwani, H.M.; Alotaibi, M.M.; Algieri, C.; Chakraborty, S. Synthesis and Characterization of Blended Cellulose Acetate Membranes. *Polymers* **2022**, *14*, 4. [CrossRef] [PubMed]
10. Ali, M.; Zafar, M.; Jamil, T.; Butt, M.T.Z. Influence of glycol additives on the structure and performance of cellulose acetate/zinc oxide blend membranes. *Desalination* **2011**, *270*, 98–104. [CrossRef]

engineering proceedings

MDPI

Proceeding Paper

Development of a Computational Tool for Maneuver Loads Estimation in Initial Design Phase for Fighter Aircraft †

Chaudry Rohan Rafiq * and Waqar Ahmed Qureshi

Department of Mechanical Engineering, University of Engineering and Technology Taxila, Punjab 47050, Pakistan
* Correspondence: rohan@aerospace.pk
† Presented at the 2nd International Conference on Advances in Mechanical Engineering (ICAME-22), Islamabad, Pakistan, 25 August 2022.

Abstract: The determination of aircraft loads in the preliminary design phase is important for the optimal design of aircraft structures. In this paper, the inflight load calculation methodology for aircraft in the preliminary design phase is formulated in the form of a computer tool developed in MATLAB/Simulink environment. In this tool (Loads Model), aircraft maneuvers are simulated, and the corresponding time histories of loads (shear force, bending moment) on the aircraft components (wing, horizontal tail, vertical tail, fuselage) are obtained, which are required for the preliminary structural design of the aircraft. The loads tool can be used to implement the design loads requirements established in "NATO RTO-45" for control configured fighter aircraft.

Keywords: loads calculation methodology; aircraft loads; load cases; inflight loads; maneuver loads; Simulink; loads model; dynamic inversion controller; loads tool development

Citation: Rafiq, C.R.; Qureshi, W.A. Development of a Computational Tool for Maneuver Loads Estimation in Initial Design Phase for Fighter Aircraft. *Eng. Proc.* **2022**, *23*, 2. https://doi.org/10.3390/engproc2022023002

Academic Editors: Mahabat Khan, M. Javed Hyder, Muhammad Irfan and Manzar Masud

Published: 20 September 2022

1. Introduction

The structural design cannot be initiated unless the loads acting on the aircraft are accurately determined. The procedures for design loads calculation for different aircraft types are readily available in the literature [1,2]. Modern military fighter aircraft are controlled by a fly-by-wire system and carefree handling technology. Hence, the determination of loads for aircraft with carefree handling capability and a load-limiting flight control system is different from conventional methods. For the design loads determination of military fighter aircraft, the NATO RTO-45 [3] specifications have been followed, which requires a loads tool to be developed for maneuver loads estimation.

A loads model (loads tool) is a platform in which the critical inflight load cases are simulated, and the loads of aircraft for different components are determined. In this paper, a loads model is developed in the Simulink environment of MATLAB. The loads model and the guidelines defined in [3] are used to determine the flight loads acting on the aircraft that are subsequently required for the preliminary design of the aircraft structure.

2. Framework of Inflight Loads Calculation Tool

In the loads model, the flight dynamics block is linked with a feedback controller to perform the maneuvers specified in different military aircraft design standards for loads estimation. The flow of the whole loads model has been defined to present a clear picture of the load's framework that is being used for aircraft loads estimation.

The flight dynamics model (FDM) provides the aircraft states (roll rate(p), pitch rate(q), yaw rate(r), angle of attack(α), sideslip(β), load factor (Nz)) time histories as a result of control inputs. The outputs of the flight dynamics model are used as inputs for the loads computation block, which calculates the loads acting on various aircraft components. The

Loads Model developed in Simulink is shown in Figure 1. The framework of the Loads Model is shown in Figure 2.

$$\begin{bmatrix}\dot{u}\\ \dot{v}\\ \dot{w}\end{bmatrix}=\frac{1}{M}\begin{bmatrix}X\\ Y\\ Z\end{bmatrix}+g\begin{bmatrix}-sin\theta\\ sin\Phi cos\theta\\ cos\Phi cos\theta\end{bmatrix}+\begin{bmatrix}rv-qw\\ pw-ru\\ qu-pv\end{bmatrix} \tag{1}$$

$$\begin{bmatrix}\dot{p}\\ \dot{q}\\ \dot{r}\end{bmatrix}=\begin{bmatrix}I_{xx} & 0 & -I_{xz}\\ 0 & I_{yy} & 0\\ -I_{zx} & 0 & I_{zz}\end{bmatrix}^{-1}+\begin{bmatrix}L+\left(I_{yy}-I_{zz}\right)qr+I_{xz}(pq)\\ M+\left(I_{zz}-I_{xx}\right)pr+I_{xz}\left(r^2-p^2\right)\\ N+\left(I_{xx}-I_{yy}\right)pq-I_{xz}qr\end{bmatrix} \tag{2}$$

$$\begin{bmatrix}\dot{\Phi}\\ \dot{\theta}\\ \dot{\psi}\end{bmatrix}=\begin{bmatrix}1 & \frac{sin\Phi sin\theta}{cos\theta} & \frac{cos\Phi sin\theta}{cos\theta}\\ 0 & cos\Phi & -sin\Phi\\ 0 & \frac{sin\Phi}{cos\theta} & cos\Phi cos\theta\end{bmatrix}\begin{bmatrix}p\\ q\\ r\end{bmatrix} \tag{3}$$

where u, v, and w are the velocities and X, Y, Z, L, M, and N are the moments about the body axes of aircraft, where Φ, θ, and ψ are the bank, elevation, and directional angle of the aircraft.

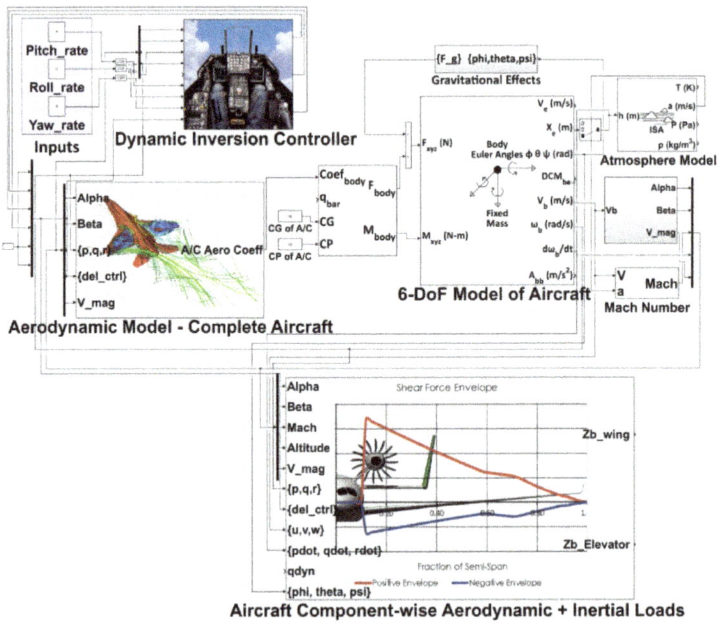

Figure 1. Loads tool developed in Simulink.

A Dynamic Inversion (DI) Controller is implemented in the Simulink Loads Model to generate the required control surfaces deflections (δe (elevator deflection), δa (aileron deflection), δr (rudder deflection)) to make the aircraft follow desired rates (p, q, r). The NDI controller has been modeled in Simulink using the guidelines outlined in [4,5].

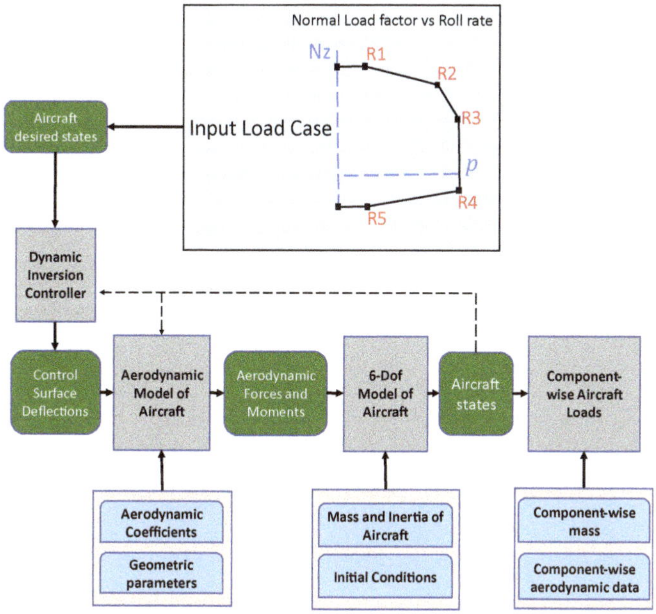

Figure 2. Loads tool framework.

Aircraft Loads (Component-Wise)

The maneuver state variable time-histories of the aircraft are used in conjunction with the component-wise aerodynamic coefficients for the calculation of component lift, drag, side force, and twisting moments. The loads equations for all the components of the aircraft are formulated using the approaches described in [1,6–8].

The equations for the lift of the wing and horizontal tail are given in Equations (4) and (5). The equation of drag is given in Equation (6), where C_L and C_D are the lift and drag coefficients, respectively.

$$L^e(t) = q_\infty S_W \left[C^e_L\left(\alpha^e(t),\ \beta^e(t),\ M,\ h\right) + C^e_{L\delta e}\left(\alpha^e(t),\ \beta^e(t),\ M,\ h\right) \delta e \right. \tag{4}$$

$$L^W(t) = q_\infty S_W \left[C^w_L\left(\alpha^W(t),\ \beta(t),\ M,\ h\right) + C^w_{L\hat{q}}\left(\alpha^W(t),\ \beta(t),\ M,\ h\right)\ \hat{q}(t)\right) \frac{\bar{c}_W}{2V_\infty} \\ + C^w_{L\delta a}\left(\alpha^W(t),\ \beta(t),\ M,\ h\right) \delta a \tag{5}$$

$$D^{e,W}(t) = q_\infty S_W \left[C^{e,w}_D\left(\alpha^{e,W}(t),\ \beta^{e,W}(t),\ M,\ h\right)\right] \tag{6}$$

where q_∞ and S_W are the dynamic pressure and wing reference area of the aircraft. The superscript represents the coefficient of the respective component. Additionally, the change in the angle of attack due to aircraft rotational rates has also been incorporated. The equation of the wing angle of attack is shown in Equation (7).

$$\alpha^W(t) = \alpha(t) + \tan^{-1}\left(\frac{p(t) y^w_{cp}}{u - r(t) y^w_{cp}}\right) \tag{7}$$

The equations for the vertical tail are also formulated using the same approach and assumptions. Finally, the aerodynamic and inertial loads are added, and resultant shear force and bending moments at the root of the respective component are obtained.

3. Results and Validation of Loads Model

3.1. Validation with DynaFlight Results

For the validation of the tool developed in this research, the trim analysis of the Loads Model developed in Simulink for a sample aircraft is compared with the results of existing aircraft loads calculation software, "DynaFlight".

The sample aircraft for which trim analysis has been performed is shown in Figure 3. The aerodynamic model of the sample aircraft and user interface of Dynaflight is shown in Figure 4. The mass of the aircraft is 18,500 kg.

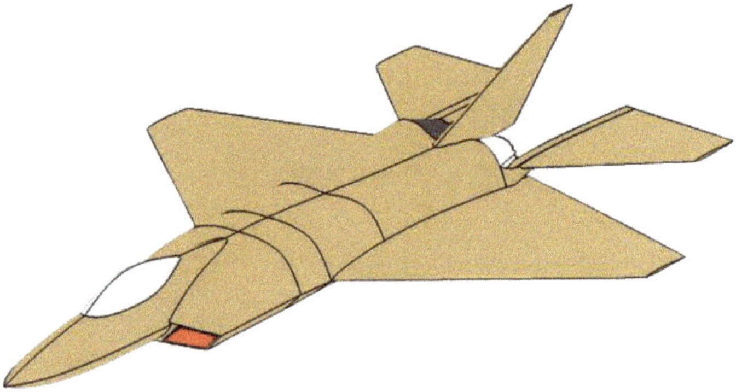

Figure 3. Aircraft model.

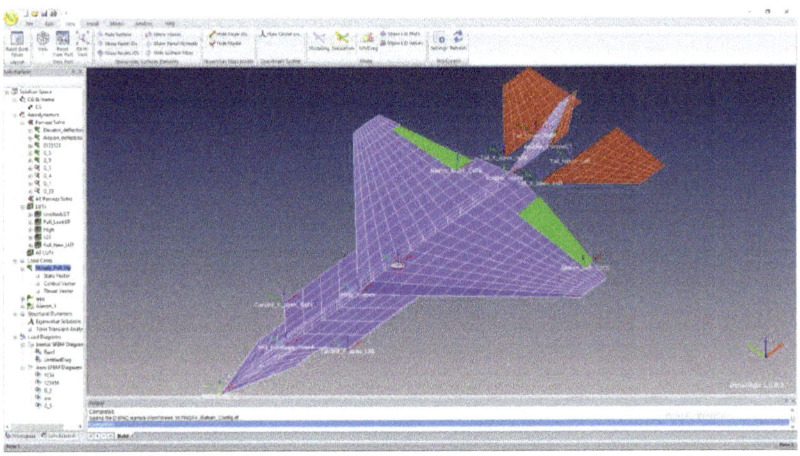

Figure 4. DynaFlight interface.

The comparison of the trim analysis for both the software for 5G pull-up and 1G level flight conditions at 0.8 Mach (M) at sea level is shown in Tables 1 and 2.

Table 1. Results for 5G pull-up maneuver.

Parameter	Simulink	DynaFlight	Difference
Angle of Attack (deg)	1.4	1.6	12.5%
H-Tail Deflection (deg)	−8.9	−9.19	3%
Shear Force on Wing (N)	651,892	658,993	1%
Shear Force on Horizontal Tail (N)	−242,342	−243,211	0.4%

Table 2. Results for 1G level flight condition.

Parameter	Simulink	DynaFlight	Difference
Angle of Attack (deg)	−1.36	−1.47	7.4%
H-Tail Deflection (deg)	−5.48	−5.57	2%
Shear Force on Wing (N)	363,436	309,395	14.8%
Shear Force on Horizontal Tail (N)	−211,619	−212,509	0.5%

For simplicity and fast computation of results, panel methods are used to obtain the aerodynamic data of the aircraft. The effects of unsteady or transient aerodynamics on wing lift due to unsteady pitching maneuvers are ignored. The aerodynamic data are obtained at discrete points, and the linear interpolation is used by the built-in Lookup tables of Simulink. For simplicity, the center of pressure (CP) is kept at 0.25 of the mean aerodynamic chord (MAC) of the lifting surface.

In longitudinal trim conditions, the load is also carried by the fuselage. The load on the fuselage is not shown in the table above. The error in the wing load for the 1G condition is larger due to the way the aerodynamic data of the individual component is stored in the Simulink model. To reduce the error, the aerodynamic data of the individual components should be obtained by taking into consideration the effects of lifting surfaces present in front of the wing. The angle of attack is a flight parameter, and the difference in the value of the angle of attack (α) could be the result of different numerical solution techniques used for the solution of flight dynamics equations in both software.

3.2. A Sample Load Case According to Flight Parameter Envelopes Approach (FPEA)

A run case for determining the maneuver loads of the sample aircraft according to the guidelines of the FPEA described in [3] is shown in this section.

The combination of the normal load factor and the roll rate serves as a critical case for the rolling maneuvers. Hence, the loads for the normal load factor coupled with the roll rate have been simulated in the loads model of Simulink, and the results are shown.

The case on the flight parameter envelope as R2 in Figure 2 has been simulated in the Simulink Loads model, and the corresponding shear force and bending moment on the wing are displayed. At R2, the Nz is "5", and the roll rate (p) is taken as "150 deg/s".

The wing and tail loads of the sample aircraft have been found using the simulation of 2 s in Figure 5. The simulation shows the time histories of aircraft flight parameters and component loading. The time history of the value of Load Factor (Nz) along with the roll rate (p) shows that the NDI controller works fine for the tracking of desired parameters, and the components' shear force and bending moments can be determined.

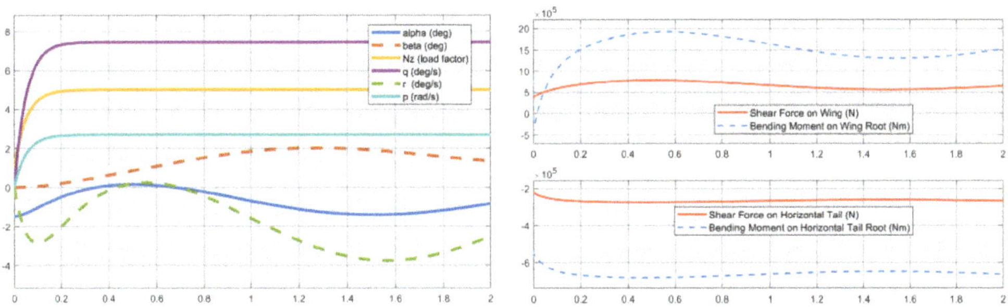

(**a**) (**b**)

Figure 5. (**a**) Angle of attack (α), sideslip (β), normal load factor (Nz), pitch rate(q), yaw rate (r), and roll rate (p) vs. time. (**b**) Shear force on wing and horizontal tail, bending moment on wing root and horizontal tail root vs. time.

4. Conclusions and Future Recommendations

The loads tool has been developed, validated, and further used to determine the loads on different components of the aircraft. Additionally, a critical load case according to the Flight Parameter Envelopes Approach (FPEA) has been simulated, and the loads on the aircraft wing and horizontal tail are determined. All the other flight parameter envelopes of the aircraft given in [3] need to be simulated for all the possible Mach numbers, altitudes, and mass states for complete loads analysis. The capability of outputting the loads time histories and having a controller for maneuver simulation makes this tool much more feasible for maneuver loads determination compared to "Dynaflight".

The flexibility effects of the aircraft structure need to be incorporated into the current loads tool. Additionally, an aerodynamic tool for the determination of aerodynamic forces and moments in the simulation run time needs to be included in the current loads determination tool to make this tool a complete standalone loads software.

Author Contributions: Conceptualization, C.R.R. and W.A.Q.; methodology, C.R.R.; software, C.R.R.; validation, C.R.R.; writing—original draft preparation, C.R.R.; writing—review and editing, W.A.Q. All authors have read and agreed to the published version of the manuscript.

Funding: This research received no external funding.

Institutional Review Board Statement: Not applicable.

Informed Consent Statement: Not applicable.

Data Availability Statement: Not applicable.

Conflicts of Interest: The authors declare no conflict of interest.

References

1. Howe, D. *Aircraft Loading and Structural Layout*; Professional Engineering Publishing: London, UK, 2004.
2. Krüger, W.; Klimmek, T. Definition of a Comprehensive Loads Process in the DLR Project ILOADS. In Proceedings of the Deutscher Luft- und Raumfahrtkongress 2016, Braunschweig, Germany, 13–15 September 2016.
3. Nato Research and Technology Organization Neuilly-Sur-Seine (France). Design Loads for Future Aircraft; 2002. Available online: https://www.sto.nato.int/publications/STO%20Technical%20Reports/RTO-TR-045/TR-045-$$ALL.pdf (accessed on 19 September 2022).
4. Vadali, R.; Majji, M.; Huff, G. *Nonlinear Adaptive Dynamic Inversion Control for Variable Stability Small Unmanned Aircraft Systems*; Texas A&M University: College Station, TX, USA, 2017.
5. Smith, S.C.; Swaroop, D.V.A.H.G.; Valasek, J.; O'neal, D. *Aircraft Control Using Nonlinear Dynamic Inversion in Conjunction with Adaptive Robust Control*; Texas A&M University: College Station, TX, USA, 2004.
6. Lomax, T.L. *Structural Loads Analysis for Commercial Transport Aircraft*; AIAA: Reston, VA, USA, 1996.
7. Niu, M.; Niu, M.C.Y. *Airframe Stress Analysis & Sizing*, 1st ed.; Hong Kong Conmilit Press Ltd.: Hongkong, 1997.
8. Wright, J.R. *Cooper Introduction to Aircraft Aeroelasticity and Loads*; John Wiley & Sons: Hoboken, NJ, USA, 2008; Volume 20.

MDPI
St. Alban-Anlage 66
4052 Basel
Switzerland
Tel. +41 61 683 77 34
Fax +41 61 302 89 18
www.mdpi.com

Engineering Proceedings Editorial Office
E-mail: engproc@mdpi.com
www.mdpi.com/journal/engproc

www.ingramcontent.com/pod-product-compliance
Ingram Content Group UK Ltd.
Pitfield, Milton Keynes, MK11 3LW, UK
UKHW060124010826
14089UKWH00006B/141

9 783036 557373